INTERATOMIC POTENTIALS

INTERATOMIC POTENTIALS

IAN M. TORRENS

Centre d'Études Nucleaires de Saclay
France

ACADEMIC PRESS
New York and London   1972

ACADEMIC PRESS, INC.
111 Fifth Avenue, New York, New York 10003

United Kingdom Edition published by
ACADEMIC PRESS, INC. (LONDON) LTD.
24/28 Oval Road, London NW1

LIBRARY OF CONGRESS CATALOG CARD NUMBER: 72-77333

PRINTED IN THE UNITED STATES OF AMERICA

Observe how system into system runs

ALEXANDER POPE

CONTENTS

Chapter III.   Interatomic Potentials Based on Thomas–Fermi Theory

Chapter IV.   Empirical Interatomic Potentials

Chapter V.   Pseudopotential Theory

Chapter VI.   **Pair Potentials Based on Pseudopotential Theory**

Chapter VII.   **Atomic Collision Theory and**
Interatomic Potentials

Chapter VIII.   **Experiments on the Scattering of Atoms and Ions**

Chapter IX.   Liquid Metal Pair Interaction Potentials

Chapter X.   The Application of Interatomic Potentials

PREFACE

The subject of the interaction between atoms is one which crosses the frontiers of many different fields of physical research. Wherever a problem is treated on an atomic scale there is a need for some knowledge of the forces which exist between the atoms. It is these forces which decide much of what we can observe in natural phenomena and a familiarity with them enables us to manipulate the raw materials we find in nature to suit our various needs. Interatomic forces are not simple because the atom itself is not an elementary particle. It is a composite body consisting of nucleus and orbital electrons which must be treated quantum mechanically. Consequently, any theory of interatomic forces or potentials must deal with the complicated problem of many-body interactions.

A large volume of literature exists in a number of different fields concerning the interaction between atoms, both at high energies where they penetrate each other until the nuclear repulsion predominates and at low energies or large interatomic separations. The scope of this literature is very wide, and few extensive reviews are available to aid the researcher new to the subject of atomic forces. To an increasing extent different fields of theoretical study call for at least an empirical knowledge of the interatomic potential and its variation with distance. In particular, with the modern large-capacity high-speed computer the treatment of physical problems numerically on an atomistic scale has become quite feasible and suitable interaction potentials between atoms are required. Even without the aid of the computer, simple analytical or statistical theories often call for an interatomic potential based on theoretical considerations or experimental data.

From the personal experience of the author, it is very difficult, when confronted with the atomic interaction problem at an early stage in a research career, to obtain a suitable interatomic potential from previous work. An elaborate literature search is usually required, and it is very easy to miss important points and to decide upon what are in some cases unsuitable forms of potential for treating the problem under consideration. Essentially, this lack of a comprehensive treatise on the subject of interatomic potentials was instrumental in the decision to write the present monograph. I have no illusions as to its role as a comprehensive work. For example, inelastic or charge–exchange atomic interactions are mentioned only very briefly, and the book might reasonably be accused of being slightly biased (for low-energy potentials) toward the metallic form of solid. Its purpose is rather to serve as an elementary introduction to the subject for those new to the field and an easy reference book for more experienced researchers. To this end I have included a considerable number of references to original articles, reviews, and books on different aspects of the subject.

Throughout the book the aim is to sacrifice involved mathematical derivations on the altar of more qualitative physical considerations. It is of course unrealistic to exclude mathematics completely, as the large number of equations bears witness, but such as there are should present little difficulty. An undergraduate course in mathematics and elementary quantum mechanics should suffice to understand any of the physical principles presented here. I apologize to the reader if it occasionally seems that I have made an equation appear from thin air. In such cases several pages of mathematics would have been required to justify its existence, and the original reference, listed at the end of the chapter, is available for consultation by those interested in probing more deeply into a particular topic.

Consistency of notation always presents a problem when gathering material from widely differing sources. One generally finds that similarly defined quantities often differ by a few physical constants, electronic charges, or powers of atomic number! In such cases I have tried to point out any differences in notation between this book and earlier original papers relevant to the particular topic.

ACKNOWLEDGMENTS

It is a great pleasure for me to acknowledge the assistance of a number of friends and colleagues during the preparation of this manuscript. Dr. Yves Adda and other members of the Département d'Etudes Metallurgiques have been sources of encouragement throughout the work. I am grateful to Professor David Lazarus, Dr. Yves Quéré, Dr. Maurice Gerl, Dr. Mark Robinson, Dr. Dirck Van Vliet, Professor Norman March, and Dr. Daniel Schiff for making valuable comments on the whole or part of the manuscript. The assistance of Messrs. Michel Bendazzoli and Pierre Truchot in the preparation and photography of a number of the figures has been very much appreciated. In addition, my thanks are due to the Department de Calcul Electronique at Saclay for the use of the computers and peripheral devices in connection with some of the tables and figures. Finally, I wish to thank all those authors who very kindly gave me permission to reproduce their figures and in many cases sent me photographs or lent me original drawings.

INTRODUCTION

The atomic theory of matter is relatively ancient, predating the actual experimental evidence for the existence of the atom by over two millenia. Following the hypotheses of earlier Greek philosophers, Lucretius, writing in the first century B.C., gave a speculative but quite lucid account of the atomic composition of matter. In this theory the interlocking between different shapes of atoms gave rise to the physical properties of materials. This idea of the physical form of the atom determining its interaction with others persisted through the period of post-Renaissance science, and it was not until the eighteenth century that the concept of an interatomic force independent of the form of the atom became current. Newton tried to explain various chemical and physical phenomena by postulating a force which was attractive at short distances but repulsive at larger separations. Although erroneous, this was the introduction of the interaction of atoms at a distance without the necessity that they touch each other. At about the same time, a remarkably clear-sighted account of the force between atoms was put forward by Boscovich. His model suggested a repulsive force at small distance followed by several

1

attraction-repulsion oscillations of the force distance curve. The zero-force points were those at which atoms combined to form stable arrangements. Considering the amount of experimental information available at that time, this was no minor achievement. Boscovich's potential shows some remarkable correspondence with very recent forms of interaction presented at various stages of this book.

Modern theories of atomic forces began with the discovery at the turn of the century of the composite nature of the atom and of the nature of the interaction between its electrically charged components. The force between atoms was revealed to be many-body in nature, made up of nucleus–nucleus, electron–nucleus and electron–electron interactions. With the development of quantum mechanics it became theoretically possible to write down a rigorous expression for the interaction of two atoms. However, the solution of the relevant equations was shown to be extremely complicated except for the simplest systems and quite substantial approximations were necessary. Although with present-day computing systems quite large advances have been made in solving the many-body problem, it is more usual to represent the interatomic force by an approximative expression. This is achieved either by approximating the theory in some way or by adjusting parameters to experimental data, or by a combination of both methods. The purpose of this book is to discuss the physical principles behind a range of such atomic interactions and to show how they may be applied to some atomic problems.

CHAPTER I

THE NATURE OF INTERATOMIC FORCES

1.1 ATOMIC INTERACTIONS IN PHYSICAL PHENOMENA

Almost all physical phenomena, discounting the world inside the atomic nucleus, may be attributed directly or indirectly to the forces between atoms. Basic concepts such as temperature and pressure, the strength of a solid or the viscosity of a liquid, as well as our own physical form and that of this book, are intimately related to the forces between atoms.

It has long been known that an atom, made up of nucleus and orbital electrons, is largely empty space, like the solar system. Therefore, the idea of an atom as a basic entity is rather nebulous. But such are the forces binding the electrons to the nucleus that except for quite high energy collisions the atom may be thought of as an entity about as penetrable, from the point of view of another atom, as a solid hard rubber ball. This is, of course, an over-simplification. In order to estimate the force acting between atoms in an ensemble or in a collision process, it is often necessary to take into account the elementary components of the atom. Thus, in the rubber ball analogy, we would have to explain why the hardness varies with the energy of a collision and why under certain circumstances two rubber balls attract each other when they are not touching.

With the normal concept of an atom made up of a central nucleus and orbital electrons, let us consider the forces intervening when it interacts with another atom. We discount any subnuclear phenomena since the forces holding together the nuclear components—neutrons, protons, and mesons— are of a completely different nature and orders of magnitude stronger than any interatomic forces. Thus the nucleus is effectively a solid body of diameter $\sim 10^{-12}$ cm with a positive charge depending on the number of protons present. If there were no orbital electrons the force between two nuclei would be Coulombic, of the form

$$F_N(r) = Z_1 Z_2 e^2 / r^2 \tag{1.1}$$

where Z_1 and Z_2 are the numbers of protons contained in the nucleus.

The nucleus + orbital electrons constitute an atom of diameter $\sim 10^{-8}$ cm, the effect of the electrons being to neutralize the charge of the nucleus. Consequently, if atoms were considered as entities there should be no forces between them at distances greater than that at which they touch. However, at separations comparable with the atomic diameter, the Coulomb interactions of the outer electrons of atom 1 with those of atom 2 and with the nucleus of atom 2 (and vice versa) have a significant effect. The two atoms no longer view each other as entities and the interaction becomes a rather complicated many-body problem. The same problem exists in an even more acute form when the electron shells of two atoms interpenetrate during an energetic collision.

To treat this many-body problem accurately for all types of atomic interaction would defy the most powerful computing systems available now or in the foreseeable future. Fortunately, a number of approximations and averaging procedures under different physical situations enable us to obtain analytical expressions or numerical values for the interatomic forces which may be applied with varying degrees of confidence to atomic problems. The greater part of this book will be devoted to the various approximations which render the many-body problem of atomic interactions less intractable.

For most purposes the force between two atoms is expressed in terms of their potential energy of interaction, or *interatomic potential*. It depends to a first approximation on the separation r between the atoms; the relation between the force $F(r)$ and the potential $V(r)$ is

$$F(r) = -(\partial/\partial r)[V(r)] \tag{1.2}$$

Strictly speaking, the potential may also depend on the relative positions of the atoms (in some types of solid or molecule), but the restriction to r-dependence is usually a good approximation.

In this chapter we shall first define in more detail the many-body problem, the potential energy of the atom, and the interaction of an ensemble of atoms. We shall then discuss the types of interatomic force or potential which apply at different atom–atom separations, and the ways in which these interactions lead to various physical phenomena.

1.2 Concept of an Atomic Potential Energy: The Many-Body Problem

Rigorously, the potential energy of an atom is the work done in bringing all components of the atom from infinity to their equilibrium positions in the atom. We shall here remain within the realm of atomic physics and exclude subnuclear phenomena. The potential energy baseline then is that of a system with the nucleus already fully constituted. Then we must calculate the work required to attach the atomic electrons to the nucleus under the force-fields of the nucleus and of each other.

The simplest example with which to introduce the concept is the semi-classical picture of the hydrogen atom. For an electron orbit of radius r, the potential energy of the atom $\phi(r)$ is the work done in bringing an electron from infinity to a distance r from the center of the nucleus under the attractive Coulomb field:

$$\phi(r) = \int_{\infty}^{r} (e^2/r^2)\, dr = -(e^2/r) \tag{1.3}$$

We may adequately describe the hydrogen atom by using this expression for $\phi(r)$ in the Schrödinger equation

$$\nabla^2 \psi + (2m/\hbar^2)(E - \phi)\psi = 0 \tag{1.4}$$

This leads to the familiar quantization of the electron states, introducing the principal, azimuthal, and magnetic quantum numbers n, l, and m.

In a larger atom, where there are several electrons, the problem rapidly becomes more involved. In the approximation of central forces, where each electron is assumed to move under the influence of the nucleus only and independent of every other electron, $\phi(r)$ would be simply given by

$$\phi(r) = \sum_{i=1}^{Z} (-Ze^2/r_i) \tag{1.5}$$

for an atom of atomic number Z. The exact solution of the Schrödinger equation would be similar to that of the hydrogen atom. The situation is unfortunately not so straightforward since the electrons are mutually repelled under a Coulomb force law, so that the actual potential energy becomes

$$\phi(r) = \sum_{i \neq j = 1}^{Z} (-(Ze^2/r_i) + e^2/2r_{ij}) \tag{1.6}$$

where r_{ij} is the distance between the ith and jth electron.

With this potential the solution of the Schrödinger equation is a rather complex problem unless the electron–electron repulsion may be considered a small perturbation on the nuclear potential term. Nevertheless, much valuable qualitative information on atomic systems may be extracted from a "hydrogenlike" treatment of the atom, neglecting electronic interactions to a first approximation.

The three quantum numbers arising from a hydrogenlike treatment of a many-electron atom are not sufficient to account for the electronic distribution among the possible energy states or to explain the experimentally observed spectral line splitting. It is necessary to introduce a fourth quantum number s which is associated with the electron spin. Although first introduced empirically by Goudsmit and Uhlenbeck [1], this fourth quantization falls naturally out of the Dirac relativistic theory.

These four quantum numbers now provide the limitation in occupation of available electron states through the *Pauli exclusion principle*, which requires that no two electrons in one atom may possess four identical quantum numbers. This law defines the electron energy distribution in the atom. Each spatial state, defined by a particular value of n, l, and m, will be occupied by at most two electrons whose spins are antiparallel. This phenomenon is responsible for the electronic structure of atoms and ions and through this explains the periodic table of the elements. The Pauli principle is also the starting point for the explanation of the chemical bond and the repulsion between the shells of bound electrons in colliding atoms or ions. It is consequently a dominant factor in the interatomic potential at small and intermediate separations.

1.3   THE POTENTIAL ENERGY OF A PAIR OF ATOMS OR IONS

In an extension of Eq. (1.6) we may now write an expression for the potential energy of a complete system of nuclei and electrons. Suppose that we have two atoms or ions of nuclear charges $Z_1 e$ and $Z_2 e$ containing a total of n electrons and separated by a distance r_{12}, and that the nuclei and electrons interact according to a Coulomb force law. Then the total potential energy is

$$\Phi(r_{12}) = \frac{1}{2} \sum_{i \neq j = 1}^{n} \frac{e^2}{r_{ij}} - \sum_{i=1}^{n} \left(\frac{Z_1 e^2}{r_{i1}} + \frac{Z_2 e^2}{r_{i2}} \right) + \frac{Z_1 Z_2 e^2}{r_{12}} \tag{1.7}$$

The first term represents the electron–electron repulsion, the second electron–nuclear interaction, and the third the repulsive energy between the two nuclei. It is evident that this is an expression of considerable complexity even for the relatively simple case of two helium atoms. However, the fact that under normal circumstances the electrons of the system have a much higher kinetic energy than the two nuclei suggests an approximation which reduces the above equation to more manageable proportions. This procedure, known as the *Born–Oppenheimer approximation* [2], consists of separating the electron and nuclear motions and treating each independently. The total wave function of the system then is made up of an electron component and a nuclear component wave function:

$$\Psi = \psi_e \, \phi_n \tag{1.8}$$

The Born–Oppenheimer approximation will be considered in greater detail in Section 2.2.

Although it will not greatly concern us in this book, since our interest lies mainly with atoms or ions, the potential energy of a system containing two *molecules* would contain additional terms involving the vibration, relative orientation, and rotation of these molecules. In an approximation of a similar nature to the Born–Oppenheimer separation, we may remove the vibration-dependent term, leaving a function which depends on the separation and relative orientation of the molecules.

The advent of high speed large capacity computing systems has been of great benefit to the calculation of potential energy of atoms and of their interactions. It is now possible to compute reliably using only the most basic approximations the atomic and molecular orbitals of atoms and simple molecules. The Schrödinger equation for nuclei and electrons can be solved and the lowest energy configuration obtained by variation. This form of approach may be used to determine the interatomic potential or to explore electron excitation or ionization processes and the nature of the chemical bond. The rate of progress in this field and in that of computer development is such that it may shortly be possible to follow in detail the behavior of the electrons and nuclei during simple chemical reactions. An elementary description of this use of the computer in the calculation of molecular orbitals has been given recently by Wahl [3].

1.4 TYPES OF INTERATOMIC FORCE

We have defined the problem of the interaction of two many-body systems in its most general terms, indicating the complexity of a rigorous treatment. Quite evidently, in all but the very simplest cases it would be both desirable and necessary to reduce the problem to that of two interacting particles. The many-body effects could be included to some appropriate degree of accuracy in the interaction potential of the two particles. This approach is of course more successful at large separations than at small values of r where some degree of "merging" of the two particles occurs. Let us now examine the different types of force which make up the interaction between two atoms or ions over a large range of separation. Both experimental phenomena and theoretical considerations suggest the separation of *short-range* and *long-range* forces, of which, for the reasons presented above, somewhat more precise information is available about the latter.

The nature of the long-range force depends on whether the system consists of neutral atoms or charged ions, or a combination of the two. The force of greatest magnitude at large separations is the Coulomb electrostatic interaction between two charged ions, assumed to be point charges:

$$V_Q(r) = -(e_1 e_2/r) \tag{1.9}$$

where e_1 and e_2 are the charges of the ions. Here and in what follows we express the force $F(r)$ in terms of the interaction potential, given for central force-fields by

$$V(r) = \int_r^{\infty} F(r)\, dr \tag{1.10}$$

If one or both of the particles is neutral, this Coulomb force is zero in the two-body approximation and the long-range interaction is greatly reduced. There remain however contributions of an inductive nature due to the presence of dipole and higher multipole moments. When for example an ion approaches a neutral atom, its effect is to attract the atomic electrons and repel the nucleus, inducing a dipole moment in the atom. The net effect is an attractive force between the now polarized atom and the ion (see Fig. 1.1(a)), which may be calculated quantum mechanically to yield a potential varying as the inverse fourth power of r. If the ion itself is replaced by a point dipole, the variation is with r^{-6}.

In a similar fashion, dispersive forces exist between neutral atoms at large separations because of the interaction between the electron distributions of the two atoms. In order to minimize the electron contribution to the potential energy of the system, any asymmetry in the electron distribution of one of the atoms leading to an instantaneous dipole moment will induce a similarly directed dipole moment in the other (see Fig. 1.1(b)). The interaction between these two dipoles results in an attractive potential varying as r^{-6}. This is known as the London–van der Waals force. Higher-order dispersive forces due to higher multipole terms also exist and give potentials varying as r^{-2n} $(n > 3)$.

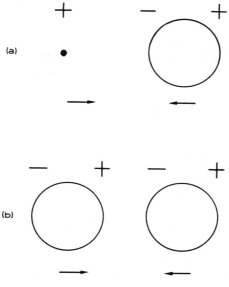

Fig. 1.1

As two atomic systems approach each other, the two-body approximation becomes less satisfactory, except in the other extreme of very high-energy collisions where the overwhelming contribution to the force is nuclear. The many-body problem has been solved for such simple systems as He–He, but the accurate quantum mechanical treatment of the interaction between heavier atoms becomes impracticable. Some quite basic approximations in the atomic model are then necessary. As has been pointed out in Section 1.2, the short-range forces arise as a consequence of the Pauli principle. They manifest themselves as either repulsive elastic forces or attractive valence forces. There exists as yet no truly reliable method of calculating from first principles the short-range interaction for atomic systems. However, approximative theoretical methods using a variety of atomic models may be combined with experimental results to yield reasonable pair interaction potentials in some ranges of atomic separation. A large part of this book is devoted to such experimental and theoretical efforts to establish the form of the short-range interaction.

1.5 THE CHEMICAL BOND

The simplest and most basic type of bond is that which unites two hydrogen atoms to form an H_2 molecule. Each atom possesses one electron in the ground state, which according to the Pauli principle can accommodate two electrons with opposing spins. In the hydrogen molecule the ground state contains two electrons with a full overlapping of spatial probability distribution but opposite spin distribution. The energy of such a molecule is lower than that of two separate atoms, leading to an attractive interatomic potential balanced at small separations by the nuclear repulsion. The formation of a hydrogen molecule through the gradual approach of two isolated atoms is illustrated by Fig. 1.2(a), reproduced from the computer-generated electron density distributions of Wahl [3]. The potential energy at various stages of this process is included in this figure.

A similar form of interaction occurs in the case of more complicated atoms so long as each atom has one electron in its outermost shell. This holds true in particular for the alkali atoms, which form molecules by sharing the valence electrons in a hydrogenlike fashion. Atoms with a single electron in their outermost shell are termed *unsaturated* units, being available for chemical bonding. The biatomic molecule is then *saturated* in the ground state, since it consists of completely filled electron states, and the bond is *covalent* or *homopolar*. These terms are not limited to molecules formed from two identical atoms, nor to bonds involving the sharing of more than one electron, although the properties exhibited by such bonds may vary from that of the simple hydrogenlike bond. Once it is a saturated unit, an atom or molecule no longer possesses a valence attractive force and will repel another like unit. An example of the interaction of two saturated units is provided by the case of two helium atoms, illustrated by Fig. 1.2(c).

A different type of bond occurs when an atom with a single outermost electron interacts with another atom lacking a single electron in order to complete its outermost shell. Then an electron is transferred between the neutral atoms, leaving two ions of unlike charge which are attracted by Coulomb forces. Such a bond is termed *ionic* or *heteropolar*, and is illustrated by the alkali halide molecules such as lithium fluoride (see Fig. 1.2(b)). It should be stressed that it is not possible to form a purely ionic bond in a molecule for reasons of symmetry. Some degree of covalency is always present. Only in crystals may the ionic bond be considered pure. Ionic bonding is not restricted to atom pairs which transfer only one electron, but the tendency to form closed shell ions is far greater when one-electron transfers are involved, and consequently the bond is very stable.

When the covalent bond of a diatomic homonuclear molecule is weak compared to that of the hydrogen molecule, we have the basis for condensation into a *metal* in aggregate form. The metallic bond is weaker than the covalent, being formed by unsaturated electrons moving freely through a lattice of atoms. It is a nonlocalized form of bonding, whose strength comes from the overlap of the electron distribution of each atom with that of its neighbors, to the extent permitted by the exclusion principle.

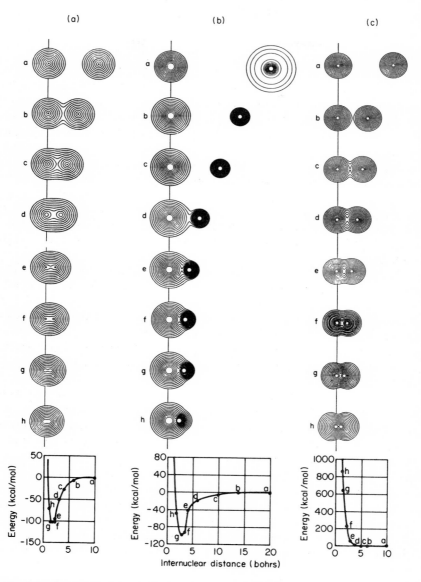

Fig. 1.2. Electron density distribution contours generated by computer solution of the Schrödinger equation for interactions in three atomic systems (a) H_2, (b) LiF, and (c) He_2. At the bottom is plotted the potential energy of interaction at the different stages of the process a–h. The separation is in atomic units. (After Wahl [3].)

We shall now consider the modifications in molecular bonding that occur when atoms aggregate macroscopically in crystalline solids.

1.6   COHESION IN CRYSTALS

When a large number of atoms are held together by chemical bonding, they usually take the form of a regular lattice whose structure is determined by the characteristics of the bonding. This is not to say that short-range interactions are solely responsible for the crystallinity of some solids. Graphite, for example, consists of layers held together by weak cohesive van der Waals interactions since there are no valence bonds between layers. Other van der Waals solids are formed by the condensation of large molecules into molecular crystals—the more complicated the molecule the greater the stability against dissociation by thermal vibration. Thus simple molecular crystals such as the rare gas solids only exist at low temperatures, whereas graphite (which may be considered a molecular crystal where each layer is a large molecule) is highly stable.

It has been mentioned in the preceding section that the bonding of Li and F in a LiF molecule is mainly ionic but retains some covalent character. When such molecules condense into a solid the single $Li^+ - F^-$ bond of the molecule gives way to multiple ionic bonds with each of the neighboring ions of opposite sign and the covalency is lost. We may consequently think of these crystals as being made up simply of positive and negative ions, which is convenient in energy calculations and other theoretical treatments (see also Appendix 3).

Atoms with a small number of valence electrons in their outermost shell and a large inner core tend to be held together in solids by nonlocalized metallic bonding. On the other hand, atoms which form molecules with strong covalent bonds, and whose ion core is much smaller than the whole atom, retain these bonds in a diversified form. In a diamond, for example, each carbon atom has a one-electron covalent link with four other atoms.

Many of the physical properties of a crystalline solid are intimately related to the type of bonding between the atoms. The electronic properties of metals arise from the overlap of electron distributions of neighboring atoms. The strength of a solid depends on the strength of the bond. Thus diamond and silicon carbide with rigid purely covalent bonds are very hard. The strongly bound ionic crystals are also rigid, whereas the metallic bond permits a small elastic deformation of the metal and the gliding of planes of atoms over one another under stress, *viz.* plastic deformation. The graphite lattice illustrates the effect of two different types of bonding as it shears easily in a direction parallel to the layers of carbon atoms owing to the feeble van der Waals bonding. Often the physical characteristics of a solid may be altered greatly by the addition of impurity atoms which either introduce extra bonds or alter the nature of the bonds existing. The classic example of this is the addition of some carbon to iron in order to form steel of much greater rigidity. The production of new types of solid with the exaggeration of certain physical properties by altering the atomic configuration and bonding has accompanied the recent rapid growth in the field of materials science.

1.7   The Relation of Physical Properties to the Interatomic Potential

The influence of atomic interactions on the macroscopic properties of a substance has already been stressed. Since these physical characteristics are sensitive to the interatomic potential it follows that we might expect to obtain some information on the latter, at least in the range of separation corresponding to equilibrium conditions, from observable physical properties. This procedure has indeed been widely used, particularly as a means of determining the values of constant parameters assigned to theoretical analytical potentials. For example, the virial coefficients of gases have been used to relate the interatomic potential to the equation of state of a gas through the medium of statistical thermodynamics. Also, the elastic constants and compressibilities of metals may be expressed relatively simply in terms of the interatomic potential, and pair correlation functions in the statistical mechanics of liquids are connected to the atomic pair interactions. We shall see illustrations of the interconnection between physical properties and the interatomic potential at various stages throughout the book.

1.8   Energetic Collisions of Atoms and Ions

We pass now from the interaction between atoms in equilibrium to that between particles whose relative velocity exceeds that of thermal motion. This brings us to the range of interatomic potential at distances less than the equilibrium separation in a solid, how much less depending on the relative energy of the collision. Consequently, some degree of closed shell interpenetration is involved, with considerable modification of the particle wave functions at the moment of impact. As previously pointed out, the theoretical approach to interpenetration is in most cases difficult if not impossible without making radical approximations. Could some information be obtained about the interatomic potential at nonequilibrium separations by observing the collision between two atoms or ions? If we know the energies and scattering angles in any collision it should be possible to find the value of the interatomic potential at the distance of closest approach. This is, of course, assuming that the fact that the collision begins and ends at infinite separation is negligible, i.e., that the potential is a rapidly varying function at the minimum separation distance.

A considerable amount of work has been devoted to the interpretation of experimental scattering data in terms of interaction potentials, both by the adjustment of a theoretical model to the experimental results and through the use of inversion procedures to obtain the numerical potential directly from the scattering cross sections.

A knowledge of the interatomic potential at small separations is of crucial importance in problems involving for example radiation damage in solids. Although much is known of the interaction of atoms at equilibrium separations in crystals, very little experimental information exists at smaller values of r. Multiple scattering processes such as the penetration of energetic charged particles in solids have been used to provide parameters for empirical interatomic potentials. By nature of the many-body problem and averaging procedures used in the theory, however, this is necessarily a very approximative method. Because of the lack of other suitable experimental approaches, the scattering of atoms and ions is a very important experimental tool for probing the potential, and scattering data provide a valuable check on theoretical models. We shall here refrain from further elaboration of this very pertinent problem, since Chapters VII and VIII are given over to a more thorough and detailed examination of the relation between atomic collisions and the interatomic potential.

1.9   THE VALIDITY OF THE TWO-BODY APPROXIMATION

Since we shall be concerned almost entirely with pairwise interatomic potentials, a fundamental question would be that of the validity of the assumption that an assembly of atoms, be it solid, liquid, or gas, may be described in terms of two-body forces between atoms. Going back to fundamental principles we would be asking whether the wave functions of the atoms in the assembly are greatly altered from their isolated values by the interaction, or whether the modification may be considered the result of a small perturbation of the free-atom function. In the latter case the two-body approximation is valid, whereas the former poses problems in the calculation of interaction potential energies. In general, atoms whose electrons are saturated are not greatly disturbed when they approach each other up to the point of interpenetration. Unsaturated units, on the other hand, have their electronic structure substantially altered when they are brought together. This would at first glance seem to limit the two-body potential model to a very narrow range of substances such as ionic crystals and van der Waals solids. However, various approximations exist for dealing with metals and to some extent covalent materials, which yield pair potentials valid under certain conditions.

As an example of such an approximation, the pseudopotential formulation relating to metals whose ion cores do not overlap in the solid provides a method of expressing the interaction of these cores by way of the intervening conduction electrons as a pair potential. This potential is valid under conditions where the conduction electron distribution is not significantly altered from that of the perfect crystal. Then the total energy of the metal may be expressed as the sum of pair interaction energies and a constant term which includes the electron kinetic energy and individual ion contributions. We shall consider the pseudopotential approach to the pair interaction in Chapters V and VI.

When we are assuming a certain analytical form of interatomic potential and wish to obtain parameter values from some experimental information it is important to bear in mind the extent to which the pairwise interaction is valid in describing the physical phenomenon involved. The simple example of the use of experimental elastic constants serves to illustrate this point. For the normal metals with small ion cores the ion–electron–ion interaction is predominant, while for noble or transition metals there is a significant repulsive force due to closed shell overlap in the solid. Nevertheless, in both cases there is a contribution to the elastic constants due to the conduction electron gas, which is volume-dependent and may not be completely described in terms of purely two-body forces.

1.10 SUMMARY

A schematic diagram of a typical two-body interatomic potential is shown in Fig. 1.3. It consists of an attractive part at large separations approaching a

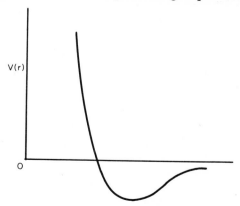

Fig. 1.3. Schematic representation of a simple interatomic two-body potential.

minimum in the region of the equilibrium separation, then becoming repulsive and increasing rapidly as r decreases further and the closed shells of electrons of the two atoms begin to overlap. The exact form of the interaction in the low energy region will differ for two atoms interacting in a crystal and in a diatomic molecule since in the former case it is influenced by the atoms in its neighborhood. Thus the equilibrium separation occurs at a different value of r for a molecule than for the atoms in a crystal. It is therefore inappropriate to use potentials obtained using molecular data in problems regarding crystals and vice versa.

Our present knowledge about the interatomic potential is relatively good at large values of r in the region of and greater than the equilibrium separation, owing to a large amount of available experimental data. Also, the theory is in quite a satisfactory state at very small separations, where nuclear scattering is predominant. There is a serious lack of information in the region between the two extremes, which is usually compensated by extrapolation of either high-energy or equilibrium potentials based on a reasonable physical model. Because of the many-body effects arising from closed shell interpenetration at these intermediate separations, theoretical models for two-body potentials are very approximative. The direct experimental determination of $V(r)$ using atom–atom scattering is in most cases difficult, but seems to be a quite promising approach if the experimental difficulties can be overcome and a sufficiently large amount of data obtained over a wide energy range.

REFERENCES

1. S. Goudsmit and G. Uhlenbeck, *Z. Phys.* **35**, 618 (1926).
2. M. Born and J. R. Oppenheimer, *Ann. Phys.* **84**, 457 (1927).
3. A. C. Wahl, *Sci. Amer.* **222** (4), 54 (1970).

CHAPTER II

.THEORETICAL MODELS OF THE ATOM

In this chapter we propose to examine some of the theories of the atom which employ various approximative methods to simplify the many-body problem and estimate, among other things, its potential energy. As elsewhere in the book, we hope to concentrate on the physical basis of the model, and omit most of the complicated mathematical development.

2.1 THE CENTRAL FIELD HYPOTHESIS

When we write down the potential energy of an atom in the form $\phi(r)$, we are assuming that its variation with distance is not directionally dependent. In fact the force field as seen by one electron is due not only to the effect of the nucleus, but also to the other electrons, and this removes to some extent its purely central nature. However, if the close approach of two electrons causes only a small deviation from the nucleus-dominated $\phi(r)$, it is a reasonable hypothesis that each electron moves in a spherically symmetrical field due to the combined effect of the nucleus and all other electrons. The constant nuclear contribution is of the order of Z times the fluctuating potential due to a nearby electron, so that the validity of the central field assumption increases with heavier atoms. In the case of a helium atom for example, the approximation is less well founded. There the electrons tend to be distributed on opposite sides of the nucleus. Thus the potential as seen by one electron depends not only on the distance from the nucleus but on the instantaneous position of the other electron. The force is then not strictly central. But for larger atoms there is a spatial averaging of the electron distribution about the nucleus, which, together with the increased nuclear charge, makes the central field hypothesis reasonably physical.

2.2   THE BORN–OPPENHEIMER APPROXIMATION

In Section 1.3 we introduced the concept of a separation of electronic and nuclear contributions to the potential energy of an atomic system for the purposes of simplification. Let us now examine this procedure in a little more detail since it represents a very important step in the theory of the atom.

Consider the Hamiltonian of a system containing N nuclei and I electrons, whose polar coordinates are $\mathbf{r}_n$ and $\mathbf{r}_i$ ($n = 1$ to N and $i = 1$ to I):

$$\mathscr{H} = -\sum_{n=1}^{N} \frac{\hbar^2}{2M_n} \frac{\partial^2}{\partial \mathbf{r}_n^{\,2}} - \sum_{i=1}^{I} \frac{\hbar^2}{2m} \frac{\partial^2}{\partial \mathbf{r}_i^{\,2}} + \frac{1}{2} \sum_{i \neq j = 1}^{I} \frac{e^2}{|\mathbf{r}_{ij}|}$$
$$+ \sum_{n=1}^{N} \sum_{i=1}^{I} \frac{Z_n e^2}{|\mathbf{r}_{ni}|} + \frac{1}{2} \sum_{n \neq l = 1}^{N} \frac{Z_n Z_l e^2}{|\mathbf{r}_{nl}|} \tag{2.1}$$

The last three terms of this equation are the different contributions to the potential energy and M_n and m are respectively the nuclear and electron masses.

Suppose that there exists a solution to the Schrödinger equation for this system using an eigenfunction Ψ composed of two independent functions, $\psi(n, \mathbf{r})$ for the electrons, which depends on the nuclear positions although the nuclei are assumed to be "frozen," and $\eta(n)$ for the nuclei:

$$\Psi = \psi(n, \mathbf{r})\eta(n) \tag{2.2}$$

Then $\psi(n, \mathbf{r})$ satisfies the electron wave equation

$$\left\{ \sum_{i=1}^{I} -\frac{\hbar^2}{2m} \frac{\partial^2}{\partial \mathbf{r}_i^{\,2}} + \sum_{i \neq j = 1}^{I} \frac{e^2}{|\mathbf{r}_{ij}|} + V(n, \mathbf{r}) \right\} \psi(n, \mathbf{r}) = E_e(n)\psi(n, \mathbf{r}) \tag{2.3}$$

The n signifies the dependence of the electron wavefunctions on the positions of the nuclei, and $V(n, \mathbf{r})$ is the electron–nucleus potential energy term of

Eq. (2.1) with the nuclei fixed. Equation (2.2) may now be substituted into the Schrödinger equation of the complete system:

$$\mathscr{H}\Psi = E\Psi \tag{2.4}$$

to give

$$\mathscr{H}\Psi = \left\{ -\sum_{n=1}^{N} \frac{\hbar^2}{2M_n} \frac{\partial^2}{\partial \mathbf{r}_n^2} + E_e(n) + W(n) \right\} \Psi \tag{2.5}$$

where $W(n)$ is the nuclear potential energy term of Eq. (2.1). Now by expansion

$$\frac{\partial^2 \Psi}{\partial \mathbf{r}_n^2} = \psi \frac{\partial^2 \eta}{\partial \mathbf{r}_n^2} + 2 \frac{\partial \psi}{\partial \mathbf{r}_n} \frac{\partial \eta}{\partial \mathbf{r}_n} + \eta \frac{\partial^2 \psi}{\partial \mathbf{r}_n^2} \tag{2.6}$$

So from Eqs. (2.2) and (2.5),

$$\mathscr{H}\Psi = \psi(n, \mathbf{r}) \left\{ -\sum_n \frac{\hbar^2}{2M_n} \frac{\partial^2}{\partial \mathbf{r}_n^2} + E_e(n) + W(n) \right\} \eta(n)$$

$$+ \left[-2 \sum_n \frac{\hbar^2}{2M_n} \frac{\partial \psi}{\partial \mathbf{r}_n} \frac{\partial \eta}{\partial \mathbf{r}_n} - \sum_n \frac{\hbar^2}{2M_n} \eta \frac{\partial^2 \psi}{\partial \mathbf{r}_n^2} \right] \tag{2.7}$$

Consider the term in square brackets, in the two cases where we have first of all free electrons, then bound electrons. For free electrons the wavefunctions are effectively independent of the nuclei, so that $\partial \psi / \partial \mathbf{r}_n$ and $\partial^2 \psi / \partial \mathbf{r}_n^2$ vanish. In the case of bound electrons we note that ψ depends on $\mathbf{r}_n$ in much the same way that it depends on $\mathbf{r}_i$ when the nuclei are fixed since it depends on $(\mathbf{r}_i - \mathbf{r}_n)$. Thus we may compare $\partial^2 \psi / \partial \mathbf{r}_n^2$ with $\partial^2 \psi / \partial \mathbf{r}_i^2$, which when multiplied by $\hbar^2 / 2m$ is a measure of the kinetic energy of one of the electrons. However, in Eq. (2.7) it is multiplied by $\hbar^2 / 2M_n$, which reduces it by a factor of $\sim 10^4$. By a similar argument it may be shown that the term involving $\partial \psi / \partial \mathbf{r}_n$ is also small compared to the first term of Eq. (2.7). In a first approximation, therefore, we may neglect the term enclosed in square brackets, leaving the equation

$$\left\{ -\sum_n \frac{\hbar^2}{2M_n} \frac{\partial^2}{\partial \mathbf{r}_n^2} + E_e(n) + W(n) \right\} \eta(n) = E\eta(n) \tag{2.8}$$

which is the Schrödinger equation of the nuclei, with the electron energy becoming a potential function for the nuclei.

In a higher-order approximation than that of Born and Oppenheimer, the contribution of the second term of Eq. (2.7) must be taken into account. This would include matrix elements giving rise to electron transitions between states as the nuclei move, i.e., the electron–phonon interaction.

2.3   THE HARTREE SELF-CONSISTENT FIELD METHOD

The basic approximations underlying the central field hypothesis prompted Hartree [1, 2] to replace the many-body interaction between one electron and all the rest of the atomic electrons by that which would result from a space-averaged charge distribution summed over those electrons and assumed to be spherically symmetrical. It then becomes a simple matter to solve the Schrödinger equation for the motion of the electron in this average potential. There are three interrelated variables in the Hartree procedure, namely the wave function of an N-electron atom, the central potential, and the charge density. These are connected by the Schrödinger equation and the laws of electrostatics. The basic problem is to arrive at a charge distribution and average potential which are self-consistent, so that this potential and the atomic wavefunction satisfy the Schrödinger equation for the system. This is accomplished by a method involving successive approximations where an initial wavefunction is chosen and the Hamiltonian of the system set up using this wavefunction. The Schrödinger equation is then solved to find a new wavefunction starting from this Hamiltonian, and if the two functions do not agree to within the specified tolerance the process is recycled starting from the new wavefunction until self-consistency is attained.

It is important for rapid convergence of the self-consistent field method, as for all successive approximation procedures, to choose a reasonable first approximation to the electron wavefunction. One way to form the wavefunction would be simply to consider it as made up of N one-electron hydrogenlike functions:

$$\Psi = \psi_1 \psi_2 \cdots \psi_n \tag{2.9}$$

The Hamiltonian of this system is

$$\mathcal{H} = -\frac{\hbar^2}{2M} \sum_i \nabla_i^2 - \sum_i \frac{Ze^2}{|\mathbf{r}_i|} + \sum_{i<j} \frac{e^2}{|\mathbf{r}_{ij}|}$$

$$= \sum_i f_i + \sum_{i<j} g_{ij} \tag{2.10}$$

where f_i and g_{ij} are respectively the one-electron and two-electron operators. This may be spherically averaged by integrating over the coordinates of the electrons:

$$\mathcal{H}_{av} = \sum_i \int \psi_i^* f_i \psi_i \, d\tau_i + \sum_{i<j} \iint \psi_i^* \psi_j^* g_{ij} \psi_i \psi_j \, d\tau_i \, d\tau_j \tag{2.11}$$

Knowing the individual ψ_i for the ith electron, which will depend on its quantum numbers, $\mathcal{H}_{av}$ may be computed and substituted into the Schrödinger equation, which may then be solved for Ψ and recycled again if necessary.

The Hartree self-consistent field, being the result of simple one-electron wavefunctions, does not fulfill the quantum mechanical requirement of complete antisymmetry in all electrons, i.e., electron exchange has been neglected. Slater [3] and Fock [4] introduced the proper antisymmetric wavefunctions into the Hartree treatment, which led to an extra exchange term in what has become known as the Hartree–Fock wave equation. This term acts as a correction on the Coulomb repulsion between an electron and the total charge distribution in the atom. Physically the exchange correction allows for the fact that an electron cannot interact with itself, but only with the charge distribution made up of the remaining $(N - 1)$ electrons.

A very simple form of static interatomic potential based on the Hartree model of the atom may be represented by the equation

$$\frac{1}{e^2} V(r) = \frac{Z_1 Z_2}{r} - Z_2 \int \frac{\rho_1(\mathbf{r}_1)\, d\mathbf{r}_1}{|\mathbf{r} + \mathbf{r}_1|} - Z_1 \int \frac{\rho_2(\mathbf{r}_2)\, d\mathbf{r}_2}{|\mathbf{r} - \mathbf{r}_2|} + e^2 \int \frac{\rho_1(\mathbf{r}_1)\rho_2(\mathbf{r}_2)\, d\mathbf{r}_1\, d\mathbf{r}_2}{|\mathbf{r} + \mathbf{r}_1 - \mathbf{r}_2|}$$

(2.12)

where $\mathbf{r}_1$ and $\mathbf{r}_2$ locate the electrons with respect to the nuclei whose separation is given by $\mathbf{r}$. The Hartree or Hartree–Fock electron densities $\rho(\mathbf{r}_1)$ and $\rho(\mathbf{r}_2)$ are found from the calculated wavefunctions.

2.4   THE THOMAS–FERMI STATISTICAL MODEL OF THE ATOM

This method used by Thomas [5] and Fermi [6] to obtain a simplified model for the many-electron atom is based on the Fermi–Dirac statistics of a free-electron gas. In its turn it is the theoretical basis of some of the most reliable empirical forms of interatomic potential at close separation. The necessary assumption for application of Fermi–Dirac statistics is that the potential energy changes very little in a volume localizing many electrons. They may then be considered the elements of a degenerate gas. We divide the space occupied by the atomic electrons into unit cells such that this assumption is valid. Then the energy states in each cell are filled subject only to the exclusion principle and to the condition that no states of positive total energy may be occupied since an electron in such a state would not be bound.

It is a familiar result of free electron theory that we may put only one energy state into a "volume" of wave number space equal to h^3. Hence we may put into this volume two electrons of opposed spin. Suppose that in an atom, at a distance r from the nucleus, all momentum space up to $p_0(r)$ is occupied. Then the number of electrons per unit volume is given by

$$\rho(r) = (2/h^3) \cdot (4/3)\pi p_0{}^3(r) = (8\pi/3h^3) p_0{}^3(r)$$

(2.13)

If $\phi(r)$ is the electrostatic potential of the atom, the maximum kinetic energy at any distance r is $-e\phi(r)$, so that the maximum occupied energy level E_0 is

$$E_0 = (p_0{}^2(r)/2m) - e\phi(r) \tag{2.14}$$

The assumption that all negative states are bound would make E_0 zero; instead we shall consider the general case where we replace E_0 by $-e\phi_0$:

$$\rho(r) = (8\pi/3h^3)(2me)^{3/2}[\phi - \phi_0]^{3/2} \tag{2.15}$$

This is the Thomas–Fermi (TF) relation between density and potential and holds for $\phi_0 < \phi$.

We now introduce self-consistency in $\phi(r)$ through the use of Poisson's equation since the potential is produced both by the nuclear charge and by the electron distribution:

$$\nabla^2\phi = -4\pi e\rho(r) \tag{2.16}$$

Eliminating $\rho(r)$ between Eqs. (2.15) and (2.16), we have

$$\nabla^2(\phi - \phi_0) = -\mu(\phi - \phi_0)^{3/2} \tag{2.17}$$

where

$$\mu = (32\pi^2 e/3h^3)(2me)^{3/2} \tag{2.18}$$

The equation is commonly converted to dimensionless form by a change of variables, of the form

$$\phi - \phi_0 = -(Ze/r)\chi(x) \tag{2.19}$$

This means that we are screening the simple Coulomb field by the function $\chi(x)$. Suppose also that

$$r = ax \tag{2.20}$$

where

$$a = \left(\frac{3}{32\pi^2}\right)^{2/3} \frac{h^2}{2me^2 Z^{1/3}} \cong \frac{0.88534 a_0}{Z^{1/3}} \tag{2.21}$$

and a_0 is the Bohr radius:

$$a_0 = h^2/4\pi^2 me^2 \tag{2.22}$$

Upon substitution of Eqs. (2.18)–(2.21) into (2.17) and a little differential algebra, the dimensionless TF equation is obtained

$$d^2\chi/dx^2 = \chi^{3/2}/x^{1/2} \tag{2.23}$$

The assumptions that the potential behaves like a simple Coulomb interaction in the extreme case as r approaches zero and that χ and χ' approach
zero as r approaches infinity, yield the boundary conditions of the TF
equation

$$\chi(0) = 1, \qquad \chi(\infty) = \chi'(\infty) = 0 \tag{2.24}$$

Methods of solution of the TF dimensionless equation will be discussed in
Section 2.7.

2.5 VARIATIONAL DERIVATION OF THE
THOMAS–FERMI ELECTRON DENSITY

The kinetic energy E_k of a system of electrons may be written in the form
of an integral involving the electron density ρ.
From Eq. (2.13);

$$E_k = \int_0^\rho (p_0{}^2/2m) \, d\rho' = c_k \rho^{5/3} \tag{2.25}$$

where

$$c_k = (3h^2/10m)(3/8\pi)^{2/3} \tag{2.26}$$

We may now use the electron density to write down the total energy of the
nucleus and electrons in an integral representation, as distinct from summation over individual electrons:

$$E = c_k \int \rho^{5/3} \, d\tau + e \int \rho\phi_n \, d\tau + \tfrac{1}{2}e \int \rho\phi_e \, d\tau \tag{2.27}$$

where

$$\phi_n = -Ze/r \tag{2.28}$$

and

$$\phi_e = e \int \rho(\mathbf{r}')/|\mathbf{r} - \mathbf{r}'| \, d\tau' \tag{2.29}$$

At this stage a Lagrangian multiplier ϕ_0 is introduced in order to fulfill the
requirement that the total number of electrons should remain equal to Z:

$$\delta[E + Ze\phi_0]_\rho = 0 \tag{2.30}$$

Varying ρ leads to an equation equivalent to (2.15):

$$\rho = (8\pi/3h^3)(2me)^{3/2}[\phi_n + \tfrac{1}{2}\phi_e - \phi_0]^{3/2} \tag{2.31}$$

We see that the Lagrangian multiplier is effectively the zero-energy reference
of the system.

2.6 INTRODUCTION OF EXCHANGE:
THE THOMAS–FERMI–DIRAC EQUATION

In the TF theory it has been assumed that the electron density distributions in the atom are independent, which is clearly contrary to the predictions of the exclusion principle. The limitation in the occupancy of a particular state to one electron will add another term to the total energy, which may be calculated by perturbation methods using the correct antisymmetrical wave functions for free electron states. The *exchange energy* thus evaluated [7] may be expressed in terms of the electron density, by the equation:

$$E_{ex} = -c_e \int \rho^{4/3} \, d\tau \tag{2.32}$$

where

$$c_e = (3e^2/4)(3/\pi)^{1/3} \tag{2.33}$$

When E_{ex} is added to the total energy expression (2.27) and the variational principle applied as before, the result is now the Thomas–Fermi–Dirac (TFD) electron density [8]:

$$\rho(r) = (8\pi/3h^3)(2me)^{3/2}[(\phi - \phi_0 + b^2)^{1/2} + b]^3 \tag{2.34}$$

where

$$b = (2me^3)^{1/2}/h \tag{2.35}$$

A change of variables analogous to that of Eqs. (2.19)–(2.21) converts to dimensionless form. Suppose

$$\phi - \phi_0 + b^2 = (Ze/r)\chi, \qquad r = ax \tag{2.36}$$

with a defined by Eq. (2.21). Then the TFD dimensionless equation is obtained, which may be compared to Eq. (2.23):

$$d^2\chi/dx^2 = x[\alpha + (\chi/x)^{1/2}]^3 \tag{2.37}$$

where

$$\alpha = 6^{1/3}/4(\pi Z)^{2/3} \tag{2.38}$$

The main difference between the TF and TFD equations is that for the latter the point of tangency of the $\chi(x)$ curve to the x-axis occurs not at infinity but at a finite value of x. The electron density $\rho(r)$ is in the TFD case no longer zero when χ is zero, but instead vanishes discontinuously at a finite value of r, the TFD boundary radius (see Fig. 2.1). Although the inclusion of the exchange correction has contracted the infinitely expansive TF electron

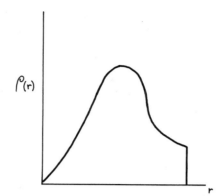

$\rho(r)$

Fig. 2.1. TFD electron density $\rho(r)$ plotted against distance r from the nucleus (schematic), showing the sharp cutoff at the TFD radius.

r

cloud, this spurious discontinuity at the TFD radius is rather unsatisfactory. One of its consequences is the disappearance at separations greater than twice the TFD atom radius of the interaction between two atoms. The electron clouds according to this theory do not overlap at all and the atoms are effectively neutral entities.

Another difference in the two theories is that the TF screening function $\chi(x)$ is a universal function depending only on x, whereas the equivalent TFD screening function is different for each value of Z and must consequently be recalculated for each atom. The principal differences between the TF and TFD atoms are summarized in Table 2-1.

TABLE 2.1

COMPARISON OF TF AND TFD ATOMS

TF atom	TFD atom
Ignores exchange, includes electron self-interaction	Includes exchange, eliminates electron self-interaction
$\rho(r) = (8\pi/3h^3)(2me)^{3/2}(\phi - \phi_0)^{3/2}$	$\rho(r) = (8\pi/3h^3)(2me)^{3/2}$ $\times [(\phi - \phi_0 + b^2)^{1/2} + b]^3$ $b = (2me^3)^{1/2}/h$
$\phi - \phi_0 = (Ze/r)\chi$ $r = ax$ $a = 0.88534a_0/Z^{1/3}$ $\chi'' = \chi^{3/2}/x^{1/2}$	$\phi - \phi_0 + b^2 = (Ze/r)\chi$ $r = ax$ $a = 0.88534a_0/Z^{1/3}$ $\chi'' = x[\alpha + (\chi/x)^{1/2}]^3$ $\alpha = 6^{1/3}/4(\pi Z)^{2/3}$
$\chi = 0$ implies $\rho = 0$ at $r = \infty$	$\chi = 0$ implies nonzero ρ and discontinuity at $r = r_0$
Screening function is universal	Screening function must be recalculated for each atom

2.7   Solution of the Thomas–Fermi Equation
for the Isolated Atom

The TF equation (2.23) has been solved by a number of investigators for the isolated neutral atom, with varying accuracy. The accurate solution is of necessity numerical, one procedure being to integrate the differential equation either outward from $x = 0$, or inward from $x = \infty$. Although $\chi(0)$ is known to be unity, one very important quantity is the value of the slope at $x = 0$, $\chi'(0)$. Early workers generally performed the outward integration for a sequence of values of $\chi'(0)$ converging to a final curve which would approach the x-axis asymptotically [9–12]. Accuracy was however difficult to achieve, since the integral was rather sensitive to the value of $\chi'(0)$ and its tangency to the x-axis at infinity was difficult to judge.

Because of the inherent instability of the outward integration, it was found to be more feasible to integrate inwards from some trial value of $\chi(x)$ at large x, this value being obtained either from an extrapolation to infinity or from an asymptotic series expansion. Such an integration has been performed by several workers [13–15] and in general leads to more accurate values of $\chi(x)$ and $\chi'(x)$.

For many applications of the statistical theory of the atom it would be desirable to have available an analytical expression which satisfies the TF equation to within a reasonable approximation. This necessity has recently diminished somewhat with the development of fast digital computers, but it is nonetheless useful to examine the different approaches to the problem which have been made. We may separate these into two broad types. The first aims to express the solution in terms of a series of simple functions such as powers or exponentials, while the second attacks the TF equation directly and attempts to put it in a form which is analytically soluble. The series approximation type of solution must in many cases be handled numerically when applied to atomic problems, which naturally limits its usefulness when the exact solution is available to a high degree of accuracy. However, thanks to asymptotic series expansions, accurate solutions have become available.

In a determination of atomic eigenvalues, Latter [16] made use of an atomic potential of the TF type, which made some allowance for the electron self-interaction:

$$\phi_1(r) = (Ze/r)\chi(x) \qquad \text{if }_{\textbf{.}}\ r > r_a$$
$$\phi_1(r) = e/r \qquad\qquad \text{if }\ r < r_a \tag{2.39}$$

The field outside the atom was thus modified to allow for the self-interaction being approximated by that of a unit charge. The screening function used in Latter's work was given in the analytical form:

$$\chi(x) = \{1 + 0.02747x^{1/2} + 1.243x - 0.1486x^{3/2} + 0.2302x^2$$
$$+ 0.007298x^{5/2} + 0.006944x^2\}^{-1} \tag{2.40}$$

This deviates from the accurate solution by less than 3%.

Other forms of series representation are of use in the actual numerical solution of the TF equation. Such is the well-known Baker series expansion [9] for small values of x ($x \leq 0.44$):

$$\chi(x) = 1 + a_2 x + a_3 x^{3/2} + a_4 x^2 + \cdots = \sum_{k=1}^{\infty} a_k x^{k/2} \tag{2.41}$$

where

$$a_2 = \chi'(0) \tag{2.42}$$

Baker series coefficients up to a_{40} have been calculated. The earlier coefficients may be found to varying degrees of accuracy in papers by Feynman et al. [11], Metropolis and Reitz [17], and Kobayashi et al. [14]. The most recent work and the highest degree of accuracy (15 significant figures) appears in a thesis by Rijnierse [18] and we reproduce in Table 2.2 his values of a_k

TABLE 2.2[a]

COEFFICIENTS a_k IN THE BAKER SERIES EXPANSION FOR $\chi(x)$

k	a_k	k	a_k
0	.100 000 000 000 000 E + 01	21	.183 630 138 770 471 E − 01
1	.000 000 000 000 000 E − 99	22	−.215 554 971 245 295 E − 01
2	−.158 807 102 260 000 E − 01	23	.257 019 114 152 481 E − 01
3	.133 333 333 333 333 E + 01	24	−.309 897 649 249 270 E − 01
4	.000 000 000 000 000 E − 99	25	.377 325 337 147 117 E − 01
5	−.635 228 409 040 000 E − 00	26	−.463 731 187 493 039 E − 01
6	.333 333 333 333 333 E − 00	27	.574 928 598 343 257 E − 01
7	.108 084 410 263 792 E − 00	28	−.718 526 562 912 179 E − 01
8	−.211 742 803 013 333 E − 00	29	.904 603 152 141 467 E − 01
9	.899 671 962 902 535 E − 01	30	−.114 659 258 089 710 E − 00
10	.144 112 547 018 389 E − 01	31	.146 244 556 793 361 E − 00
11	−.271 286 107 057 753 E − 01	32	−.187 619 238 525 825 E − 00
12	−.295 055 008 478 292 E − 03	33	.242 007 418 156 681 E − 00
13	.173 443 041 642 727 E − 01	34	−.313 745 267 564 561 E − 00
14	−.167 709 556 143 035 E − 01	35	.408 678 305 314 600 E − 00
15	.112 763 141 431 646 E − 01	36	−.534 704 559 574 435 E − 00
16	−.911 028 066 991 997 E − 02	37	.702 518 484 493 754 E − 00
17	.102 762 683 449 552 E − 01	38	−.926 631 598 151 114 E − 00
18	−.123 240 451 261 030 E − 01	39	.122 677 504 660 947 E + 01
19	.141 442 924 276 242 E − 01	40	−.162 983 000 387 084 E + 01
20	−.159 860 960 349 775 E − 01		

[a] From Rijnierse [18].

of Eq. (2.41) for $k = 0$ to 40. The Baker series converges rapidly for small x but the convergence becomes rather poor as x approaches unity.

Another asymptotic expansion, this time for large x, was the result of a development of the Sommerfeld approximation (see next section). It was given by Coulson and March [19], and takes the form

$$\chi(x) = (144/x^3) \sum_{k=0}^{\infty} c_k (Fx^\lambda)^k \tag{2.43}$$

This series converges for $x \geq 4.75$, although accuracy is hindered by very slow convergence until x becomes greater than 10. Again we reproduce the figures given by Rijnierse [18] for $k = 0$ to 30 in Table 2.3. For the isolated

TABLE 2.3[a]

COEFFICIENTS c_k IN THE ASYMPTOTIC EXPANSION OF $\chi(x)$ FOR LARGE VALUES OF x

k	c_k	k	c_k
0	.100 000 000 000 000 E $+$ 01	16	.243 263 216 206 774 E $-$ 06
1	$-$.100 000 000 000 000 E $+$ 01	17	$-$.691 031 290 629 697 E $-$ 07
2	.625 697 497 782 349 E $-$ 00	18	.202 144 402 461 023 E $-$ 07
3	$-$.313 386 115 073 309 E $-$ 00	19	$-$.586 883 726 659 920 E $-$ 08
4	.137 391 276 719 371 E $-$ 00	20	.169 228 967 471 091 E $-$ 08
5	$-$.550 834 346 641 491 E $-$ 01	21	$-$.484 947 296 184 278 E $-$ 09
6	.207 072 584 991 917 E $-$ 01	22	.138 179 140 746 420 E $-$ 09
7	$-$.741 452 947 849 571 E $-$ 02	23	$-$.391 670 654 111 897 E $-$ 10
8	.255 553 116 794 870 E $-$ 02	24	.110 486 378 825 038 E $-$ 10
9	$-$.854 165 377 806 924 E $-$ 03	25	$-$.310 287 206 325 100 E $-$ 11
10	.278 373 839 349 473 E $-$ 03	26	.867 815 397 770 538 E $-$ 12
11	$-$.888 226 094 136 591 E $-$ 04	27	$-$.241 782 703 567 386 E $-$ 12
12	.278 359 915 878 395 E $-$ 04	28	.671 228 958 979 403 E $-$ 13
13	$-$.858 949 083 219 027 E $-$ 05	29	$-$.185 723 018 657 859 E $-$ 13
14	.261 505 871 330 370 E $-$ 05	30	.512 274 393 418 862 E $-$ 14
15	$-$.786 798 261 946 265 E $-$ 06		

[a] From Rijnierse [18].

atom $F = 13.270\,973\,848$ and $\lambda = -0.772\,001\,872\,6$. By introducing a new independent variable,

$$u = \beta Fx^\lambda/(1 + \beta Fx^\lambda), \qquad \beta = 0.250\,762\,378\,7 \tag{2.44}$$

Rijnierse obtained modifications of the series (2.43) in the form

$$\chi(x) = (144/x^3) \sum_{k=0}^{\infty} \gamma_k u^k, \qquad \chi'(x) = (144/x^4) \sum_{k=0}^{\infty} \gamma_k' u^k \tag{2.45}$$

with values of γ_k and γ_k' again calculated for $k = 1$ to 30. Convergence was greatly improved by this change of variable, enabling χ and χ' to be calculated to an accuracy of 12 significant figures for $x \geq 1$ and to 6 figures for $x \geq 0.6$.

By matching this series at $x = 1$ to a modified Taylor series and requiring analytic continuation, Rijnierse obtained by an iterative method an accurate value of the initial slope a_2 and of the parameters of the large-x asymptotic series. We reproduce in Appendix 1 the range of values of $\chi(x)$ and $\chi'(x)$ between $x = 0$ and $x = 1000$ given by Rijnierse to 8 significant figures precision. Inasmuch as it avoids any integration procedures, this work represents an algebraic solution of the TF equation, although of course it involves the use of computing machinery. The method of reconstituting asymptotic series solutions at large and small x so that their ranges of convergence overlap is ingenious and has led to the most accurate values of the TF screening function and its first derivative so far available. There is no reason why the inward integration should not lead to a similar degree of accuracy and confirm the results of Rijnierse, given the computing power presently available. Of course such a high degree of precision is for most purposes unnecessary, when we consider the approximations involved in the basic model.

2.8 ANALYTICAL APPROXIMATIONS
TO THE THOMAS–FERMI SCREENING FUNCTION

We pass now to the more simple forms of approximate analytical solution, which may be more feasible for application to various atomic problems where a high degree of accuracy is not demanded.

Perhaps the earliest and best-known of these is the Sommerfeld asymptotic form [20]:

$$\chi(x) = \{1 + (x/a)^\lambda\}^{-c} \tag{2.46}$$

where the constants a, λ, and c were chosen so that $\chi(0) = 1$, $\chi(\infty) = 0$.

$$a = 144^{1/3}, \qquad c\lambda = 3 \tag{2.47}$$

By fitting this equation for large x, Sommerfeld arrived at the values $\lambda = 0.772$, $c = 3.886$, so that his form was finally:

$$\chi(x) = \{1 + (x/12^{2/3})^{0.772}\}^{-3.886} \tag{2.48}$$

In later work, March [21] normalized the radial momentum distribution function using the Sommerfeld approximation and modified the value of λ to 0.8034. In a further refinement of March's investigation, Umeda [22] again modified λ to 0.8371 improving the agreement with numerical results. We note that the series expansion (2.43) is derived from the March version of the Sommerfeld approximation.

Somewhat simpler and more approximative than the Sommerfeld form were those proposed by Lindhard *et al.* [23]:

$$\chi(x) = 1 - x/(3 + x^2)^{1/2} \tag{2.49}$$

and

$$\chi(x) = 1 - 1/2x \tag{2.50}$$

Another very simple equation was that suggested by Kerner [24]:

$$\chi(x) = 1/(1 + Bx), \qquad B = 1.3501 \tag{2.51}$$

The parameter B was later adjusted by Umeda [22] to 1.3679.

With the Baker series expansion for small x and the Coulson–March–Sommerfeld series for large x, it was desirable to obtain an analytical approximation suited to the intermediate region. Brinkman [25] noticed that the TF equation could be written:

$$\chi''(x) = (x\chi)^{1/2}\chi/x \tag{2.52}$$

Brinkman's approximation was then to treat $(x\chi)^{1/2}$ as a constant since it was a slowly varying function—zero at both $x = 0$ and $x = \infty$. Then:

$$\chi''(x) \simeq A^2\chi/x \tag{2.53}$$

This now has the solution

$$\chi(x) = Cx^{1/2}K_1(2Ax^{1/2}) \tag{2.54}$$

where C is an arbitrary constant and K_1 a modified Bessel function of the second kind.

A second transformation due to Tietz [26] was similarly based on the concept of a slowly varying $(x\chi)^{1/2}$:

$$\chi'' = (x\chi)^{-1/2}\chi^2 \simeq B^2\chi^2 \tag{2.55}$$

Then

$$\chi(x) = [1 + (B/6^{1/2}) \cdot x]^{-2} \tag{2.56}$$

Umeda and Kobayashi [27] investigated the approximations of Brinkman and of Tietz by studying the variation of the generalized function $x^\sigma\chi_0(x)$, where σ was a variable parameter and χ_0 the exact TF function. When substituted into the Brinkman and Tietz forms this would yield for Eq. (2.53)

$$\chi''(x) \simeq (x^{0.6462}\chi)^{1/2}\chi/x^{0.8231} \tag{2.57}$$

and for Eq. (2.55)

$$\chi''(x) \simeq (x^{0.6462}\chi)^{-1/1.2924}\chi^{2.274} \tag{2.58}$$

In fact these values of the constants are quite close to the Brinkman and Tietz indices, so the latter are good approximations. This fact was confirmed by some further development in which Umeda and Kobayashi showed that if the TF equation were written in the generalized Brinkman and Tietz forms:

$$\chi'' = A^2 \chi / x^\alpha \tag{2.59}$$

or

$$\chi'' = B^2 \chi^\beta \tag{2.60}$$

then the only possible integer for α is 1, and the solution of Eq. (2.60) would be in the form:

$$\chi(x) = (1 + Bx)^{-n} \tag{2.61}$$

where n must be less than 3. Thus the only integral values of n are 1 and 2, giving respectively the Kerner and Tietz forms.

A number of workers have proposed analytical solutions involving exponentials or combinations of powers and exponentials. Two rather complicated exponential forms were given by Rozental [28]:

$$\chi(x) = 0.7345e^{-0.562x} + 0.2655e^{-3.392x}$$
$$\chi(x) = 0.255e^{-0.246x} + 0.581e^{-0.947x} + 0.164e^{-4.356x} \tag{2.62}$$

The difficulty in handling these forms analytically is quite evident. Molière [29] also gave an approximation in the form of three exponentials, although his equation was a little more tractable (see also Section 4.11):

$$\chi(x) = 7pe^{-qx} + 11pe^{-4qx} + 2pe^{-20qx} \tag{2.63}$$

In this equation $p = 0.05$ and $q = 0.3$.

Another equation very similar to those of Rozental and Molière was proposed by Csavinsky [30]:

$$\chi(x) = \{0.7111e^{-0.175x} + 0.2889e^{-1.6625x}\}^2 \tag{2.64}$$

This upon squaring again comes out as the sum of three exponentials.

A single parameter exponential function was proposed by Roberts [31]:

$$\chi(x) = (1 + \gamma x^{1/2}) \exp(-\gamma x^{1/2}) \tag{2.65}$$

In a variational principle treatment of this function, Anderson, Arthurs, and Robinson [32] obtained the optimum value of γ as 1.7822 and found by a similar examination of the Csavinsky equation (2.64) that the single parameter function (2.65) was a better approximation to the accurate $\chi(x)$.

The final approximation which we shall mention here is a recent one due to Wedepohl [33], who noted that in the range of x between 0.3 and 16, a quite accurate description of $\chi(x)$ is obtained by the equation

$$\chi(x) = 317x \exp(-6.62x^{1/4}) \tag{2.66}$$

The main disadvantage of this expression is that $\chi(0) = 0$ and the equation becomes rapidly erroneous as x drops below the lower limit of 0.3.

In Fig. 2.2 we compare some curves of $\chi(x)$ using two of the analytical forms listed in Table 2.4 with the numerical values of Kobayashi *et al.* [14].

TABLE 2.4

ANALYTICAL SOLUTIONS TO THE TF EQUATION

Author	Analytical expression for (x)
1. Sommerfeld [20]	$\left[1 + \left(\dfrac{x}{12^{2/3}}\right)^{\lambda}\right]^{-3/\lambda}$ $\lambda = 0.772$ (Sommerfeld [20]) $\lambda = 0.8034$ (March [21]) $\lambda = 0.8371$ (Umeda [22])
2. Kerner [24]	$(1 + Bx)^{-1}$ $B = 1.3501$ (Kerner [24]) $B = 1.3679$ (Umeda [22])
3. Brinkman [25]	$Cx^{1/2}K_1(2Ax^{1/2})$
4. Tietz [26]	$(1 + (B/6^{1/2}) \cdot x)^{-2}$
5. Rozental [28]	$0.7345e^{-0.562x} + 0.2655e^{-3.392x}$
6. Rozental [28]	$0.255e^{-.0246x} + 0.581e^{-0.947x} + 0.164e^{-4.356x}$
7. Molière [29]	$0.35e^{-0.3x} + 0.55e^{-1.2x} + 0.10e^{-6.0x}$
8. Csavinsky [30]	$(0.7111e^{-0.175x} + 0.2889e^{-1.6625x})^2$
9. Roberts [31, 32]	$(1 + 1.7822x^{1/2}) \exp[-1.7822x^{1/2}]$
10. Wedepohl [33]	$317x \exp[-6.62x^{1/4}]$
11. Lindhard [23]	$1 - x/(3 + x^2)^{1/2}$
12. Lindhard [23]	$1 - 1/2x$

There is evidently no lack of variety of analytical approximation to the Thomas–Fermi screening function. The choice of which equation to use in a given problem will usually depend on a compromise between the accuracy required and the analytical simplicity of the approximation.

2.9 CONCLUSION

We have used a large proportion of this chapter to describe in outline the TF theory of the isolated atom and the work which has been carried out concerning the solution of the TF equation. The reason for this is that while the TF model is by no means the most accurate theoretical image of the atom, it possesses the distinct advantage of simplicity combined with increasing validity for heavier atoms. In addition it provides an initial hypothesis for some of the more reliable theoretical interatomic potentials at close separations. We shall describe some of these potentials in the next chapter,

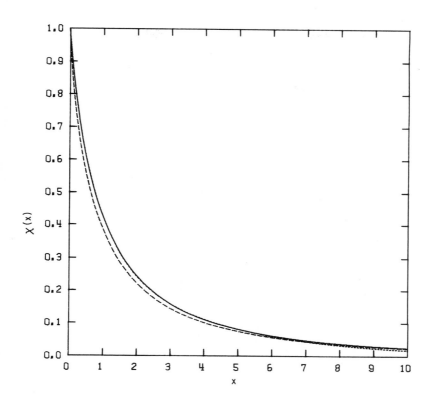

Fig. 2.2. Comparison of the numerical solution of Kobayashi *et al.* [14] with some analytical approximations to the Thomas–Fermi screening function: (——) Kobayshi *et al.*; [14]; (- - - -) Molière [29]; (– – –) Sommerfeld [20].

but to illustrate the capability we might write down immediately a crude interaction potential between two atoms of atomic numbers Z_1 and Z_2 in the form of a screened Coulomb potential using χ as the screening function: From Eqs. (2.19)–(2.21)

$$V(r) = (Z_1 Z_2 e^2/r)\chi(r/a) \tag{2.67}$$

where a is now the TF screening radius for the collision between the two atoms:

$$a = 0.88534 a_0/Z_{\text{eff}}^{1/3} \tag{2.68}$$

Z_{eff} is the effective charge number in the interaction of two unlike atoms. There exist a number of approximations for Z_{eff} as we shall see in the next chapter, but we might write down a very simple mean value:

$$Z_{\text{eff}} = (Z_1^{1/2} + Z_2^{1/2})^2 \tag{2.69}$$

Since it is impossible in one chapter to achieve a complete picture of the TF atom, we have of necessity concentrated on those aspects of the theory which are most pertinent to the development of an interatomic potential. Further information on TF theory may be obtained from the many publications cited in the text, as well as from the book by Gombas [34] and a review by March [35].

REFERENCES

1. D. R. Hartree, *Proc. Cambridge Philos. Soc.* **24**, 111 (1928).
2. D. R. Hartree, "The Calculation of Atomic Structures." Wiley, New York, 1955.
3. J. C. Slater, *Phys. Rev.* **35**, 210 (1950).
4. V. Fock, *Z. Phys.* **61**, 126 (1930).
5. L. H. Thomas, *Proc. Cambridge Philos. Soc.* **23**, 542 (1927).
6. E. Fermi, *Z. Phys.* **48**, 73 (1928).
7. F. Bloch, *Z. Phys.* **57**, 545 (1929).
8. P. A. M. Dirac, *Proc. Cambridge Philos. Soc.* **26**, 376 (1930).
9. E. B. Baker, *Phys. Rev.* **36**, 630 (1930).
10. J. C. Slater and H. M. Krutter, *Phys. Rev.* **47**, 559 (1935).
11. R. P. Feynman, N. Metropolis, and E. Teller, *Phys. Rev.* **73**, 1561 (1949).
12. C. Miranda, *Mem. Reale Accad. Ital.* **5**, 285 (1934).
13. V. Bush and S. H. Caldwell, *Phys. Rev.* **38**, 1898 (1931).
14. S. Kobayashi, T. Matsukuma, S. Nagai, and K. Umeda, *J. Phys. Soc. Japan.* **10**, 759 (1955).
15. M. T. Robinson, private communication (1970).
16. R. Latter, *Phys. Rev.* **99**, 510 (1955).
17. N. Metropolis and H. Reitz, *J. Chem. Phys.* **19**, 555 (1951).
18. P. J. Rijnierse, *in* "Studies in Radiation Effects in Solids" (G. J. Dienes and L. T. Chadderton, eds.), Vol. 5. Gordon & Breach, New York (in press).
19. C. A. Coulson and N. H. March, *Proc. Phys. Soc.* **A63**, 367 (1950).
20. A. Sommerfeld, *Z. Phys.* **78**, 283 (1932).
21. N. H. March, *Proc. Cambridge Philos. Soc.* **46**, 356 (1950).
22. K. Umeda, *J. Phys. Soc. Japan.* **9**, 290 (1954).
23. J. Lindhard, V. Nielson, and M. Scharff, *Kgl. Dansk. Vid. Selsk. Mat. Fys. Medd.* **36**, No. 10 (1968).
24. E. H. Kerner, *Phys. Rev.* **83**, 71 (1951).
25. H. C. Brinkman, *Physica* **20**, 44 (1954).
26. T. Tietz, *J. Chem. Phys.* **22**, 2094 (1954).
27. K. Umeda and S. E. Kobayashi, *J. Phys. Soc. Japan.* **10**, 749 (1955).
28. S. Rozental, *Z. Phys.* **98**, 742 (1936).
29. G. Molière, *Z. Naturforsch.* **2a**, 133 (1947).
30. P. Csavinsky, *Phys. Rev.* **166**, 53 (1968).
31. R. E. Roberts, Univ. Wisconsin Theoret. Phys. Inst. Rep. WIS-TCI (1967).
32. N. Anderson, A. M. Arthurs, and P. D. Robinson, *Nuovo Cimento* **57B**, 523 (1968).
33. P. T. Wedepohl, *Proc. Phys. Soc.* **1**, 307 (1968).
34. P. Gombas, "Die Statische Theorie des Atoms und ihre Anwendungen." Springer-Verlag, London and New York, 1952.
35. N. H. March, *Advan. Phys.* **6**, 1 (1957).

CHAPTER III

INTERATOMIC POTENTIALS BASED ON THOMAS–FERMI THEORY

3.1 INTRODUCTION

We have seen in Chapter II how the TF statistical theory of the atom produces an electronic screening function $\chi(r/a)$ upon solution of the TF dimensionless equation. We also noted briefly that this function may be used in a simple-minded interaction potential between a pair of TF atoms (see Eq. (2.67)). In this chapter we shall examine more closely such a potential and those resulting from more elaborate treatments of the two-center situation on the TF and TFD models.

It will become evident that care must be exercised in the interpretation of the physical significance of TF and TFD atoms, with reference to their two-body interaction. We remember that the electron density of the TF atom extends to infinity, so that any pair potential must be overestimated at large separations. The TFD atom, on the other hand, has a sharp electron density cutoff at a finite radius, leading to the disappearance of the interaction potential for atom separations beyond twice this radius—again a rather artificial effect. Because of these long-range effects the TF and TFD interatomic potentials are most reliable for small separations, typically less than 1 Å.

3.2   The Firsov Thomas–Fermi Pair Potential

The earliest consideration of the two-center TF problem was by Firsov, who in two quite separate treatments evolved two-body potentials of the general form of Eq. (2.67). His earlier derivation [1] assumed the approximation for the total electron energy in an atom [1a]:

$$E_{el} \simeq -0.77(4\pi^2 m e^4/h^2)Z^{7/3} \qquad (3.1)$$

Then the energy difference between two atoms entirely overlapped and entirely separate is given by the expression

$$E_{el}(r)_{r\to 0} \simeq -0.77(4\pi^2 m e^4/h^2)\{(Z_1 + Z_2)^{7/3} - Z_1^{7/3} - Z_2^{7/3}\} \qquad (3.2)$$

The term in brackets could be approximated by the formula

$$(Z_1 + Z_2)^{7/3} - Z_1^{7/3} - Z_2^{7/3} \simeq \tfrac{7}{3} Z_1 Z_2 (Z_1 + Z_2)^{1/3} \qquad (3.3)$$

Thus:

$$E_{el}(r) \simeq -1.8(4\pi^2 m e^4/h^2)Z_1 Z_2 (Z_1 + Z_2)^{1/3} \qquad (3.4)$$

The interaction potential energy $V(r)$ of the two atoms may be written as the sum of the nuclear and electronic energies:

$$V(r) = Z_1 Z_2 e^2/r + E_{el}(r) - \Delta E_{el}(r) \qquad (3.5)$$

Firsov applied first-order perturbation theory to calculate the change in energy $\Delta E_{el}(r)$ of the electrons when the two nuclei $Z_1 e$ and $Z_2 e$ draw apart by a distance r. This he postulated was equivalent in the first order to the change in energy of an atom whose nucleus is $(Z_1 + Z_2)e$ when a charge $+Z_2 e$ is placed at r with $-Z_2 e$ at the center. This is the average potential difference of the electrons between zero and r:

$$\Delta E_{el}(r) \simeq Z_2 e[\phi_1(0) - \phi_1(r)] \qquad (3.6)$$

where $\phi_1(r)$ is the TF potential of the electrons, defined by Eq. (2.19), less the Coulomb potential of the nucleus:

$$\phi_1(r) = \{(Z_1 + Z_2)e/r\}[\chi(r/a) - 1] \qquad (3.7)$$

the screening radius a being defined by the equation

$$a \simeq 0.885 h^2/4\pi^2 m e^2 (Z_1 + Z_2)^{1/3} = 0.885 a_0/(Z_1 + Z_2)^{1/3} \qquad (3.8)$$

By expanding $\chi(r/a)$ about the origin and neglecting all but the first two terms, Firsov found that his $E_{el}(r)$ of Eq. (3.5) was approximately canceled by the $\phi_1(0)$ term of $\Delta E_{el}(r)$, as would be expected from perturbation theory. His expression was asymmetric in Z_1 and Z_2, a consequence of removing the charge $Z_2 e$ from the center of the atom and carrying the calculation to first order only. This could be easily remedied by adding or subtracting a second-order term. This done, the Coulomb contributions in $V(r)$ disappeared by subtraction, leaving

$$V(r) = (Z_1 Z_2 e^2/r)\chi(r/a) \tag{3.9}$$

Thus a rather crude first-order perturbation approach to the two-center problem nevertheless leads to an expression of the appropriate screened Coulombic form.

3.3   THE FIRSOV VARIATIONAL DERIVATION OF THE THOMAS–FERMI POTENTIAL

Firsov's second interatomic potential derivation [2] was a little more fundamental in its initial assumptions, setting out this time from the total Hamiltonian of a system of nuclei and electrons (see Eq. (2.27)). At any particular separation r of two atoms it may be reasonably assumed that the electron distributions will be distorted so that this total energy $\mathscr{H}$ is a minimum for this separation. For two atoms,

$$\mathscr{H} = e\left\{ c_k \int \rho^{5/3}\, d\tau - \int \sum \frac{Z_i}{r_i} \rho\, d\tau + \frac{1}{2}\iint \frac{\rho(r)\rho(r')\, d\tau\, d\tau'}{|\mathbf{r}-\mathbf{r}'|} \right\} \tag{3.10}$$

where r_i are the distances from the origin to the nuclei. The procedure was to minimize $\mathscr{H}$ with respect to the electron density $\rho(r)$, which upon simplification leads to the expression for minimal energy:

$$\mathscr{H}_0 = (c_k/6)\int \rho_0^{5/3}\, d\tau - \tfrac{1}{2}\int \sum (Z_i/r_i)\rho_0\, d\tau \tag{3.11}$$

This then provided a lower bound for the Hamiltonian of the system. In itself this is inadequate to define the interaction energy accurately, so Firsov next introduced a functional $\mathscr{H}_1$ of a function f which was related to the electron density ρ. Upon variation of $\mathscr{H}_1$ with respect to f, $\mathscr{H}_1$ was found to be maximum when $f = f_0$, given by

$$f_0 = \sum (Z_i/r_i) - \tfrac{5}{3}c_k \rho_0^{2/3} \tag{3.12}$$

Then the minimal of $\mathscr{H}_1$ was identical to the expression (3.11) for $\mathscr{H}_0$. Thus he established an upper and lower bound for $\mathscr{H}$, although we note that no physical reality may be attached to the functional $\mathcal{H}_1$ except in the case where $f = f_0$.

The procedure in the two-center case is to replace the overall electron density by the sum of extremals of the two densities $\rho_{01}(r_1)$ and $\rho_{02}(r_2)$, found by minimizing $\mathcal{H}$ for each atom separately:

$$\rho(r) = \rho_{01}(r) + \rho_{02}(r) \tag{3.13}$$

and by replacing f by the sum of the extremals of f for each atom:

$$f = f_{01} + f_{02} \tag{3.14}$$

These represent approximations with respect to the general treatment outlined above for a system of nuclei and electrons. Consequently the value of $\mathcal{H}$ calculated using (3.13) will now differ from that of $\mathcal{H}_1$ using (3.14). By comparison of $\mathcal{H}$ and $\mathcal{H}_1$, Firsov was able to calculate the upper limit to the error made in assuming the extremal density summation of Eq. (3.12) and calculating $\mathcal{H}_0$ on that basis. This was in the region of 8%, which meant that if $\mathcal{H}_0$ were assumed to be the mean of $\mathcal{H}$ and $\mathcal{H}_1$, then

$$\mathcal{H}_0 \simeq \tfrac{1}{2}(\mathcal{H} + \mathcal{H}_1) \tag{3.15}$$

The error in $\mathcal{H}_0$ was less than $\tfrac{1}{2}|\mathcal{H} - \mathcal{H}_1|$, in this case $\sim 4\%$. The maximum error occurred for atomic interactions where ρ_{01} and ρ_{02} were close in value, and the error dropped rapidly as the values of ρ_{01} and ρ_{02} differed.

It is relatively simple to express the interatomic potential $V(r)$ in terms of these expressions for the Hamiltonians. Thus

$$V(r) = Z_1 Z_2 e^2/r + \mathcal{H} - \mathcal{H}(\infty) \tag{3.16}$$
$$V_1(r) = Z_1 Z_2 e^2/r + \mathcal{H}_1 - \mathcal{H}_1(\infty) \tag{3.17}$$

where $\mathcal{H}(\infty)$ and $\mathcal{H}_1(\infty)$ are the expressions for two entirely separate atoms. These expressions may be evaluated using the summation technique for ρ and f of Eqs. (3.13) and (3.14), and consequently the limits for $V(r)$ may be obtained. Firsov, however, wished to express the potential if possible in simpler terms somewhat similar to those of his earlier treatment (see Eq. (3.9)):

$$V_0(r) = (Z_1 Z_2 e^2/r)\chi(r/a_{12}) \tag{3.18}$$

where a_{12} is a screening radius symmetrical in Z_1 and Z_2:

$$a_{12} = 0.885 a_0/\mathscr{F}(Z_1, Z_2) \tag{3.19}$$

He tried different forms of $\mathscr{F}(Z_1, Z_2)$, including the original one:

$$\mathscr{F}(Z_1, Z_2) = \begin{cases} (Z_1 + Z_2)^{1/3} \\ (Z_1^{2/3} + Z_2^{2/3})^{1/2} \\ (Z_1^{1/2} + Z_2^{1/2})^{2/3} \end{cases} \tag{3.20}$$

For each of these and for different ratios Z_2/Z_1, the $V_0(r)$ was evaluated over a range of r and compared with the equivalent extremal values $V(r)$ and $V_1(r)$. Best agreement was found for the third form of $\mathscr{F}(Z_1, Z_2)$. Then $V_0(r)$ differed from $V(r)$ by less than 20% over a range of variation of r/a_{12} from 0 to 10, which represents, depending on Z, a value of r in the region of 1 Å. At distances exceeding this, agreement is meaningless, since calculation on the basis of the statistical model loses its validity.

Thus Firsov finally presented as his two-body interatomic potential based on TF theory, the form

$$V(r) = (Z_1 Z_2 e^2/r)\chi\{(Z_1^{1/2} + Z_2^{1/2})^{2/3}(r/0.885a_0)\} \qquad (3.21)$$

valid in the range $r \lesssim 1$ Å. The Firsov potentials for copper and argon are traced in Fig. 3.1. The most notable feature is the very slow decay of these potentials as r increases, which limits their range of application.

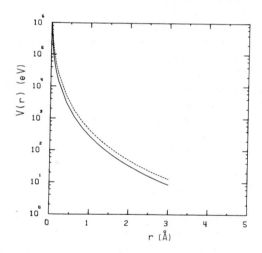

Fig. 3.1. Firsov potential for copper and argon: (———) copper; (– – – –) argon.

3.4 SOLUTION OF THE THOMAS–FERMI EQUATION
FOR THE DIATOMIC MOLECULE

Townsend and Handler [3] made a numerical study of the two-center TF problem in application to the neutral homonuclear diatomic molecule. They calculated the total energy of a molecule made up of two atoms of atomic number Z, separated by a distance $2r$, solving the TF equation by a finite difference relaxation procedure. In calculating this energy they included an electron exchange contribution. Subtracting the energy of two isolated atoms

from the total energy of the molecule, they gave the interaction energy or
interatomic potential as

$$V(r) = Z^2 e^2/r + E_{TF}(r) - E_{TF}(\infty) + E_a(r) - E_a(\infty) \qquad (3.22)$$

The results of Townsend and Handler minus the exchange contribution E_a
agree well with the interatomic potential of Firsov for the same pair of
atoms. In a sense this might be considered to be an improvement on the
Firsov model, although its basis is not so simple and it requires numerical
procedures to calculate the potential.

3.5   VARIATIONAL DERIVATION OF THE
THOMAS–FERMI–DIRAC INTERATOMIC POTENTIAL

Abrahamson and co-workers attempted to extend the maximal–minimal
principle used by Firsov to the TFD model of the atom, including the
exchange contribution. As was indicated in Chapter II, in the TFD model
the dimensionless equation was no longer universal, so that the interatomic
potential had to be numerically calculated for each pair of atoms. After the
manner of Firsov, Abrahamson et al. [4] wrote down the Hamiltonian $\mathscr{H}$ for
the system, including this time the exchange term in $\rho^{4/3}$ and found $\mathscr{H}_0$ by
minimization. Thus $\mathscr{H}_1$ was again a functional of a slightly modified function
f, and again its maximum, together with the $\mathscr{H}_0$, limited the total energy.

When applied to the two-center TFD problem, this treatment yielded an
interatomic potential of the form

$$V(r) = \frac{1}{2} \frac{Z_1 Z_2 e^2}{r} \left\{ \chi\left(\frac{Z_1^{1/3} r}{a}\right) + \chi\left(\frac{Z_2^{1/3} r}{a}\right) \right\} + \overline{\Lambda} - C_z \qquad (3.23)$$

where $\overline{\Lambda}$ is a function of ρ_{01} and ρ_{02} similar to that of Firsov, except that it
includes terms in $\rho^{4/3}$ to account for exchange. Here C_z is a parameter which
depends on Z_1 and Z_2 and is constant for a given pair of ions and χ is the
TFD screening function. Abrahamson et al. predicted that the potential
represented by Eq. (3.23) would become somewhat inaccurate when r ex-
ceeded the smaller of the two TFD electron density cutoff radii, and would
be inapplicable for r greater than the sum of the two radii, where, according
to the TFD model, there is no further interaction.

In later work, Abrahamson [5] used the above expression for $V(r)$ slightly
modified to calculate the interatomic potentials between homonuclear rare
gas atom pairs at distances ranging from 0.01 Å to about 6 Å. His purpose
was to compare these with other forms of interatomic potential and in particular with those derived from experimental atom–atom scattering results
(see Chapter VIII). The C_z parameter of Eq. (3.23) was dropped in this later

work, since it was held formally responsible for maintaining the electron density unrealistically high near the TFD cutoff radius, and was at the same time negligibly small inside the atom. The interatomic potentials of Abrahamson for He–He, Ne–Ne and Ar–Ar interactions are shown in Fig. 3.2. They appeared to be in only order-of-magnitude agreement with experimental results. Values of the Abrahamson rare gas potentials over a range of r are reproduced in Table 3.1.

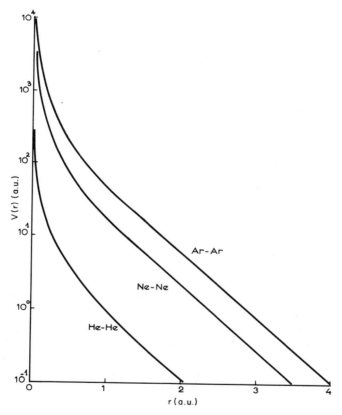

Fig. 3.2. TFD interatomic potentials of Abrahamson for He–He, Ne–Ne, and Ar–Ar interactions in atomic units (a.u.).

The work was subsequently extended to a number of heteronuclear pairs of the rare gas type [6], and a recent publication of Abrahamson [7] has attempted to overcome the nonuniversal aspect of the TFD pair interaction by fitting a Born–Mayer type of analytical potential:

$$V(r) = Ae^{-Br} \tag{3.24}$$

TABLE 3.1

TFD Interatomic Potentials of Abrahamson for Rare Gas Atoms (in Atomic Units)[a,b]

$r(a_0)$ \\ Z:	2	10	18	36	54	86
0.01	391.91	9 690.0	31 206	123 720	276 620	697 150
0.03	125.31	3 039.3	9 690.2	37 908	83 813	208 450
0.06	59.155	1 362.9	4 407.9	16 921.	37 001	90 535
0.1	33.073	747.48	2 360.7	8 881.1	19 105	46 067
0.3	7.999 5	166.40	491.01	1 714.1	3 535.6	8 001.8
0.6	2.599 7	49.404	138.75	453.15	897.32	1 922.8
1.0	0.908 24	16.229	43.807	136.51	259.12	533.61
1.5	0.303 89	5.407 8	14.298	42.843	79.906	159.95
2.0	0.107 11	2.053 5	5.411 5	16.081	29.594	58.428
2.5	0.035 695	0.799 62	2.145 9	6.452 2	11.863	23.438
3.0	0.011 982	0.297 11	0.836 33	2.582 9	4.815 2	9.516 2
3.5	(0.003 9)[b]	0.100 78	0.298 79	0.970 50	1.850 5	3.742 9
4.0	(0.001 2)	0.040 756	0.101 95	0.327 82	(0.76)	1.325 7
4.5	(0.000 42)	(0.015)	(0.038)	0.141 92	(0.30)	(0.57)
5.0	(0.000 14)	(0.005 6)	(0.014)	(0.055)	(0.12)	(0.22)
5.5	—	(0.002 1)	(0.005 2)	(0.022)	(0.044)	(0.083)
6.0	—	—	(0.001 9)	(0.008 3)	(0.018)	(0.032)
8.0	—	—	—	—	(0.000 44)	—
r_b	3.321 0	4.050 7	4.281 8	4.527 5	4.657 7	4.795 2

[a] After Abrahamson [5].
[b] Extrapolated values are enclosed in parentheses.

This is feasible because, as we note from Fig. 3.2, the TFD potential is quite linear on a semilogarithmic plot over a considerable range of r, typically from about $1.5a_0$ to $3.5a_0$, although extrapolation of the analytical curve to $\sim 6a_0$ still produced better than order-of-magnitude agreement with the full TFD potential. In this way Abrahamson was able to list the constants A and B for homonuclear pair interactions of all atoms from $Z = 2$ to $Z = 102$. This list is reproduced in Table 3.2. Using these values, a combination rule could be used to calculate all possible heteronuclear pair interactions:

$$V_{12} \simeq (V_{11}V_{22})^{1/2} \tag{3.25}$$

The theoretical basis for this combination rule in a more general form has been discussed recently by Smith [7a]. Using an atomic distortion model to describe the atoms during the collision, he finds that the mean BM parameters A and B are given by

$$(A_{12}B_{12})^{2/B_{12}} = (A_{11}B_{11})^{1/B_{11}} + (A_{22}B_{22})^{1/B_{22}} \tag{3.25a}$$

This is equivalent to Eq. (3.25) only in the case where $B_{11} = B_{22}$. However, since the variation of the BM exponential parameter B is usually quite small (see Table 3.2), the use of Eq. (3.25) as a combination rule introduces very little error in comparison with other sources of uncertainty.

Thus a simple analytical expression has been made available for over 5000 atom–atom interactions. This expression is equivalent over a reasonable range of r to that arising from the two-center solution of the TFD statistical model of the atom. Such an expression would be extremely desirable for many applications, were it not for the fact that its reliability, and indeed that of the whole Abrahamson variational treatment of the TFD problem, has been questioned by Günther [8].

Abrahamson and co-workers, in their formulation of the maximum principle, did not allow the finite boundary of the TFD electron density to be varied in the maximization. Their atomic radius was assumed to remain constant during the interaction, equal to the normal TFD radius. Günther included this condition in a reformulation of the maximal principle, and found that the extremal of the modified Abrahamson functional did not represent a maximum with respect to surface variation of the two atoms, but rather had a cubic form of behavior. Thus the maximal principle could not be used with surface variation to obtain a limit to the total energy. Günther showed that if the correct TFD radius were known from the beginning of the calculation and retained in the variation, the extremal value of the maximal principal did in fact prove to be a maximum. It would not however be known in most cases of practical interest. The unjustified constant TFD radius assumption was further shown by Günther to lead in the two-center case to an error in the development of $\mathcal{H}_1$, adding to the discrepancy in the final result.

In a convincing comparison of TFD interaction energies for two argon atoms, Günther demonstrated that the Abrahamson potential overestimated the interatomic potential by quite a serious margin. In the absence of a reliable method of calculating the exact TFD interaction, he obtained an upper limit using the results of Townsend and Handler [3] (see Section 3.4), whose TF potential without exchange agreed well with the work of Firsov. This upper limit $V'_{\text{TFD}}(r)$ was realized using the exact TFD energy and exchange energy of the isolated atoms:

$$V'_{\text{TFD}}(r) = V_{\text{TH}}(r) + E_{\text{TF}}(\infty) + E_a(\infty) - E_{\text{TFD}}(\infty) \tag{3.26}$$

where $V_{\text{TH}}(r)$ is the Townsend and Handler interaction potential given by Eq. (3.22). We have effectively

$$V'_{\text{TFD}}(r) = Z^2 e^2/r + E_{\text{TF}}(r) + E_a(r) - E_{\text{TFD}}(\infty) \tag{3.27}$$

TABLE 3.2

NUMERICAL VALUES OF BORN–MAYER PARAMETERS A AND B FOR
NEUTRAL GROUND-STATE TFD ATOMS WITH $Z = 2$ TO $Z = 105^a$

Atomic number, Z	Chemical symbol	A		B		ε^b (%)
		(e^2/a_0)	(eV)	(a_0^{-1})	(Å^{-1})	
2	He	8.6047	234.13	2.207 79	4.172 17	4.1
3	Li	16.109	438.33	2.120 81	4.007 80	4.2
4	Be	24.599	699.34	2.059 04	3.891 07	4.7
5	B	35.606	968.84	2.027 71	3.831 87	4.1
6	C	48.367	1316.1	2.015 92	3.809 59	4.5
7	N	62.840	1709.9	2.009 00	3.796 51	5.1
8	O	78.771	2143.4	2.004 74	3.788 46	5.0
9	F	96.267	2619.4	2.002 29	3.783 83	5.1
10	Ne	114.72	3121.5	1.999 54	3.778 63	5.6
11	Na	134.56	3661.4	1.998 99	3.777 59	5.8
12	Mg	140.72	3829.0	1.956 94	3.698 13	4.9
13	Al	157.85	4295.1	1.946 81	3.678 99	4.2
14	Si	186.39	5071.7	1.958 88	3.701 80	4.3
15	P	205.67	5596.3	1.951 37	3.687 60	3.9
16	S	222.99	6067.6	1.940 26	3.666 61	3.7
17	Cl	235.64	6411.8	1.924 49	3.636 81	3.6
18	Ar	255.82	6960.9	1.919 01	3.626 45	3.6
19	K	277.95	7563.0	1.916 34	3.621 41	3.6
20	Ca	298.57	8124.1	1.910 44	3.610 26	3.4
21	Sc	319.62	8696.9	1.905 44	3.600 81	3.2
22	Ti	343.72	9352.6	1.904 02	3.598 12	3.2
23	V	366.70	9977.9	1.900 77	3.591 98	3.2
24	Cr	389.77	10 606	1.897 94	3.586 63	3.1
25	Mn	414.53	11 279	1.896 07	3.583 10	2.9
26	Fe	438.46	11 931	1.893 00	3.577 30	3.0
27	Co	463.35	12 608	1.891 16	3.573 82	3.0
28	Ni	487.72	13 271	1.888 18	3.568 19	2.7
29	Cu	511.53	13 919	1.884 57	3.561 37	2.7
30	Zn	539.78	14 687	1.884 24	3.560 74	2.8
31	Ga	564.68	15 365	1.880 99	3.554 60	2.8
32	Ge	590.33	16 063	1.879 03	3.550 90	2.8
33	As	618.26	16 823	1.878 33	3.549 58	2.7
34	Se	645.35	17 560	1.876 88	3.546 84	2.8
35	Br	672.43	18 297	1.874 95	3.543 19	2.9
36	Kr	703.70	19 148	1.875 20	3.543 66	2.8
37	Rb	731.90	19 915	1.873 73	3.540 88	2.9
38	Sr	763.87	20 785	1.873 95	3.541 30	2.8
39	Y	792.10	21 553	1.872 49	3.538 54	2.8
40	Zr	819.91	22 310	1.870 42	3.534 63	2.9

TABLE 3.2   (*continued*)

Atomic number, Z	Chemical symbol	A		B		ε^b
		(e^2/a_0)	(eV)	(a_0^{-1})	(Å^{-1})	(%)
41	Nb	850.89	23 153	1.869 90	3.533 65	2.9
42	Mo	881.65	23 990	1.869 26	3.532 44	3.0
43	Tc	911.85	24 811	1.864 89	3.524 18	3.1
44	Ru	941.72	25 624	1.867 23	3.528 60	3.1
45	Rh	972.66	26 466	1.866 43	3.527 09	3.1
46	Pd	1005.2	27 352	1.865 99	3.526 26	3.0
47	Ag	1040.7	28 318	1.866 57	3.527 35	3.1
48	Cd	1071.1	29 145	1.865 21	3.524 78	3.2
49	In	1102.6	30 002	1.864 16	3.522 80	3.3
50	Sn	1140.2	31 025	1.865 10	3.524 58	3.4
51	Sb	1171.4	31 874	1.863 82	3.522 16	3.5
52	Te	1199.4	32 636	1.861 36	3.517 51	3.5
53	I	1231.2	33 501	1.860 20	3.515 32	3.6
54	Xe	1265.2	34 426	1.859 80	3.514 56	3.5
55	Cs	1298.9	35 343	1.859 01	3.513 07	3.4
56	Ba	1336.4	36 363	1.859 56	3.514 11	3.5
57	La	1370.2	37 283	1.858 91	3.512 88	3.5
58	Ce	1403.3	38 184	1.857 96	3.511 08	3.5
59	Pr	1436.0	39 074	1.856 79	3.508 87	3.6
60	Nd	1471.5	40 040	1.856 64	3.508 59	3.6
61	Pm	1508.4	41 044	1.856 51	3.508 34	3.7
62	Sm	1544.3	42 020	1.856 21	3.507 78	3.7
63	Eu	1580.5	43 005	1.856 06	3.507 49	3.8
64	Gd	1615.9	43 969	1.855 71	3.506 83	3.7
65	Tb	1651.4	44 935	1.855 12	3.505 72	3.8
66	Dy	1688.7	45 950	1.855 13	3.505 73	3.8
67	Ho	1724.9	46 935	1.854 65	3.504 83	3.8
68	Er	1765.1	48 028	1.854 93	3.505 36	3.6
69	Tm	1801.4	49 016	1.854 49	3.504 52	3.7
70	Yb	1842.3	50 129	1.854 89	3.505 28	3.8
71	Lu	1879.8	51 149	1.854 59	3.504 71	3.8
72	Hf	1914.1	52 083	1.853 44	3.502 54	3.8
73	Ta	1952.7	53 133	1.853 46	3.502 58	3.9
74	W	1991.2	54 181	1.853 00	3.501 71	4.0
75	Re	2028.1	55 185	1.852 31	3.500 41	4.0
76	Os	2067.9	56 268	1.852 40	3.500 58	4.0
77	Ir	2104.6	57 266	1.851 90	3.499 63	4.0
78	Pt	2144.2	58 344	1.851 91	3.499 65	4.0
79	Au	2185.7	59 473	1.852 04	3.499 89	4.1
80	Hg	2230.0	60 678	1.852 54	3.500 84	4.1

TABLE 3.2.   (*continued*)

Atomic number, Z	Chemical symbol	A		B		ε^b (%)
		(e^2/a_0)	(eV)	(a_0^{-1})	(Å^{-1})	
81	Tl	2270.7	61 786	1.852 46	3.500 69	4.1
82	Pb	2311.8	62 904	1.852 58	3.500 92	4.2
83	Bi	2349.2	63 922	1.851 64	3.499 14	4.2
84	Po	2395.9	65 192	1.852 24	1.500 27	4.1
85	At	2434.6	66 246	1.851 73	3.499 31	4.1
86	Rn	2476.5	67 386	1.851 63	3.499 12	4.2
87	Fr	2517.4	68 499	1.851 37	3.498 63	4.2
88	Ra	2623.0	71 372	1.863 68	3.521 89	5.7
89	Ac	2665.6	72 531	1.863 60	3.521 74	5.6
90	Th	2716.9	73 927	1.864 70	3.523 82	5.3
91	Pa	2761.1	75 130	1.864 80	3.524 01	5.2
92	U	2806.6	76 368	1.865 19	3.524 75	5.0
93	Np	2850.8	77 570	1.865 18	3.524 73	4.7
94	Pu	2895.1	78 776	1.865 32	3.524 99	4.5
95	Am	2940.2	80 003	1.865 44	3.525 22	4.3
96	Cm	2981.6	81 129	1.864 87	3.524 14	4.2
97	Bk	3027.6	82 381	1.865 11	3.524 59	4.2
98	Cf	3072.8	83 611	1.865 06	3.524 50	4.1
99	Es	3115.2	84 765	1.864 56	3.523 55	4.2
100	Fm	3160.0	85 984	1.864 61	3.523 65	4.2
101	Md	3207.8	87 284	1.864 85	3.524 10	4.2
102	No	3253.6	88 531	1.864 95	3.524 29	4.2
103	Lw	3300.2	89 798	1.865 11	3.524 59	4.2
104	—	3346.4	91 056	1.864 99	3.524 37	4.3
105	—	3393.9	92 348	1.865 25	3.524 86	4.3

[a] After Abrahamson [7].
[b] ε is the magnitude of the maximum percent error for each fit.

This represents an upper bound since $E_{TF}(r)$ and $E_a(r)$ will be greater than the exact $E_{TFD}(r)$ found by minimization. It may not be considered a good potential since $V'_{TFD}(\infty)$ is nonzero, but it is nevertheless useful as a limiting potential for the purposes of comparison. In Table 3.3, reproduced from Günther's work, the different potentials are compared for four values of interatomic separation r. There are the TF potentials of Townsend and Handler omitting and including exchange, the upper bound for the TFD potential, and the TFD potential according to Abrahamson, interpolated from Table 3.1. All potentials are for Ar–Ar interactions.

TABLE 3.3

ARGON–ARGON POTENTIAL IN ATOMIC UNITS[a]

r/a_0	0.076314	0.76314	1.52629	3.81572
V_{TF}[b]	3062.3	59.730	10.321	0.605
V_{TFD}[b]	3037.8	53.883	8.315	0.353
V' = upper bound for exact TFD potential	3039.3	55.4	9.8	1.8
V_{TFD}[c]	3222	80.57	13.54	0.152
Minimum error in Abrahamson potential in %	5.7	31	28	—

[a] After Gunther [8].
[b] After Townsend and Handler [3].
[c] After Abrahamson [5].

It may be seen from this table that in the intermediate range of r, the Abrahamson potential is considerably higher than that calculated from the Townsend–Handler work. At small r the nuclear repulsion becomes predominant and the difference is less noticeable. The table also shows that the true TFD potential lies below the TF potential over the whole range of r, instead of being above it at intermediate separations as found by Abrahamson. This is in accord with the results of Townsend and Handler including exchange, and with those of a separate study by Günther [9] on the basis of mutual penetration of the shell structure of the two interacting atoms.

3.6 ANALYTICAL THOMAS–FERMI TWO-BODY POTENTIALS

To complete the present chapter we would draw attention to the two-body potentials which make use of analytical approximations to the accurate TF screening function $\chi(x)$. These are simplifications of the Firsov potential and can use the same screening length a. They belong more appropriately to the next chapter, where we shall discuss them in relation to other forms of empirical interatomic potential.

In this chapter some rather complicated mathematical arguments have been radically pruned, and the methods have been presented only in outline. The author apologizes for any lack of clarity which results from this type of presentation. The alternative was to reproduce *in toto* the mathematical development of the authors in question, which is inappropriate when the original article is available for consultation by the interested reader.

REFERENCES

1. O. B. Firsov, *Dokl. Akad. Nauk USSR* **91**, 515 (1953).
1a. P. Gombas, "Die Statische Theorie des Atoms und ihre Anwendungen." Springer-Verlag, London and New York, 1952.
2. O. B. Firsov, *Zh. Eksperim. Teor. Fyz.* **33**, 696 (1957) [*English transl.*: *Sov. Phys. JETP* **6**, 534 (1958)].
3. J. R. Townsend and G. S. Handler, *J. Chem. Phys.* **36**, 3325 (1962).
4. A. A. Abrahamson, R. D. Hatcher, and G. H. Vineyard, *Phys. Rev.* **121**, 159 (1961).
5. A. A. Abrahamson, *Phys. Rev.* **130**, 693 (1963).
6. A. A. Abrahamson, *Phys. Rev.* **133**, A990 (1964).
7. A. A. Abrahamson, *Phys. Rev.* **178**, 76 (1969).
7a. F. T. Smith, *Phys. Rev. A* **5**, 1708 (1972).
8. K. Günther, *Ann. Phys.* **14**, 296 (1964).
9. K. Günther, *Kernenergie* **7**, 443 (1964).

CHAPTER IV

EMPIRICAL INTERATOMIC POTENTIALS

4.1 INTRODUCTION

It is perhaps misleading to use the word "empirical" in this chapter heading, in view of its usual slightly pejorative connotation. In fact the present state of theoretical knowledge of the subject is such that these potentials often present a more realistic view of atomic interactions than potentials derived exclusively and usually at great pain from purely theoretical considerations which are themselves often approximative in nature. Empirical atomic interactions are in most cases based on a simple analytical expression which may or may not be justifiable from theory, and which contains one or more parameters adjusted to an experimental situation. Their purpose is to facilitate the analytical treatment of atomic problems such as crystal properties, radiation damage studies or molecular interactions. In recent years the application of computer techniques to these problems has permitted the use of more complicated empirical potentials based on more widely varied experimental information.

By strict definition almost all forms of interatomic potential which exist at present are empirical, due to the approximations necessary to overcome the many-body problem involved in the interaction. However, we shall here interpret the term "empirical potential" as meaning a fairly simple, mathematically tractable, analytical expression for the pairwise interaction between two atoms or ions. We have already encountered a few such potentials constructed from analytical approximations to the screening function (see Section 3.6). Let us first consider some of the most elementary forms of expression.

4.2 Simple Models

The simplest concept of an atom is that it takes the form of a rigid, impenetrable, undeformable sphere. It then possesses a potential function of the form (Fig. 4.1(a)):

$$V(r) = \begin{cases} \infty, & r < \sigma \\ 0, & r > \sigma \end{cases} \tag{4.1}$$

The parameter σ represents the radius of the atom. The basic advantage of this potential is its simplicity for the purposes of calculation. It has been widely used as a first approximation in theoretical investigations where some form of interatomic potential is necessary and where quantitative accuracy is of secondary importance to a conceptual view of the phenomenon under study. One example of its use is to be found in calculations involving multiple atomic displacement cascades in a solid which occur as a result of incident high energy radiation. Here the hard sphere potential permits an analytical study of the problem leading to order-of-magnitude estimates of the number and energy distribution of displaced atoms per high energy primary knocked-on atom. It is possible to improve the validity of this potential by making the hard sphere radius σ depend on the energy of the collision.

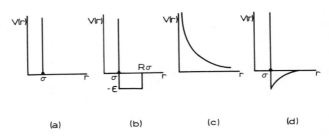

Fig. 4.1. Simple models for the interatomic potential. (a) Hard-sphere; (b) square-well; (c) Born; (d) Sutherland.

A modification of the hard-sphere potential which attempts to allow for the attractive part of the interatomic forces is the square-well potential (Fig. 4.1(b)):

$$V(r) = \begin{cases} \infty, & r < \sigma \\ -\varepsilon, & \sigma < r < R\sigma \\ 0, & r > R\sigma \end{cases} \tag{4.2}$$

In this model the rigid sphere is surrounded by an attractive shell of depth ε and thickness $\sigma(R - 1)$. It has the advantage over the simpler model that there are two extra parameters ε and R, which facilitate an adjustment to experimental results, always an important factor in the use of any empirical potential.

Instead of a "square" type of potential, we could assume that the atoms are not entirely hard but repel each other with a force which varies as an inverse power of distance:

$$V(r) = \lambda/r^n \tag{4.3}$$

Usually n is somewhere between 9 and 15 (Fig. 4.1(c)). Known as the Born potential [1], this is the basis for several other more sophisticated empirical potentials. The atoms are theoretically point centers of repulsion, although the high value of n and the constant λ give them effective radii which depend on the energy of the interaction. This potential has the advantage of easy differentiability, which is often useful in calculations where forces between atoms come into play.

Another modification of the rigid sphere model takes into account the attractive force by allowing the spheres to attract each other according to an inverse power law (Fig. 4.1(d)):

$$V(r) = \begin{cases} \infty, & r < \sigma \\ -(\lambda/r^n), & r > \sigma \end{cases} \tag{4.4}$$

In this case n is normally about 5 or 6. This is known as the Sutherland potential [2], and combines mathematical simplicity with a more realistic physical model.

In this section we have considered only the interaction between neutral atoms. In the case of charged particles it would be necessary to add a Coulomb attraction or repulsion.

4.3   The Buckingham Potentials

Buckingham suggested a form of potential which included the induced dipole-induced dipole and quadrupole–quadrupole interactions, and expressed the repulsive part by an exponential term rather than a power dependence [3]:

$$V(r) = Ae^{-Br} - \lambda/r^6 - \lambda'/r^8 \qquad (4.5)$$

This form is somewhat more difficult to handle mathematically owing to the mixing of exponentials and powers, even though it is a more realistic potential from a theoretical point of view. It has, however, the rather unfortunate property of being negatively infinite at zero separation.

A variation of this potential known as the Buckingham–Corner potential eliminated the unrealistic behavior at the origin by postulating the following rather complicated form [4]:

$$V(r) = \begin{cases} A \exp\left\{-\alpha\left(\dfrac{r}{r_m}\right)\right\} - \left(\dfrac{\lambda}{r^6} + \dfrac{\lambda'}{r^8}\right) \exp\left\{-4\left(\dfrac{r_m}{r} - 1\right)^3\right\}, & r < r_m \\[3mm] A \exp\left\{-\alpha\left(\dfrac{r}{r_m}\right)\right\} - \left(\dfrac{\lambda}{r^6} + \dfrac{\lambda'}{r^8}\right), & r \geq r_m \end{cases} \qquad (4.6)$$

where

$$\begin{aligned} A &= \{-\varepsilon + (1 + \beta)(\lambda/r_m^{\,6})\}e^\alpha \\ \lambda &= \varepsilon\alpha r_m^{\,6}/\{\alpha(1 + \beta) - 6 - 8\beta\} \\ \lambda' &= \beta r_m^{\,2}\lambda \end{aligned} \qquad (4.7)$$

Here ε is the depth of the energy minimum and r_m is the corresponding value of r. The steepness of the exponential is measured by α, and β is the ratio of the inverse 8th to 6th power contributions at $r = r_m$. There are four independent constant parameters which may be adjusted to experimental data. The main disadvantage of this potential is its complicated functional form, which makes it rather intractable for analytical purposes.

A simpler related form, which drops the inverse 8th power term, is known as the modified Buckingham (or "exponential-6") potential

$$V(r) = \frac{\varepsilon}{1 - 6/\alpha}\left\{\frac{6}{\alpha}\exp\left[\alpha\left(1 - \frac{r}{r_m}\right)\right] - \left(\frac{r_m}{r}\right)^6\right\} \qquad (4.8)$$

The constants have similar significance to those of the Buckingham–Corner potential. This potential has a spurious maximum at separation $r_{\max}$ given by the equation

$$\left(\frac{r_{\max}}{r}\right)^7 \exp\left\{\alpha\left(1 - \frac{r_{\max}}{r_m}\right)\right\} = 1 \qquad (4.9)$$

The problem may be solved by simply assuming that for $r \leq r_{\max}$ the atom becomes a hard sphere:

$$V(r) = \infty, \qquad r \leq r_{\max} \qquad (4.10)$$

The value of r_{max} is usually sufficiently small that such an approximation is acceptable. The effect of the induced dipole-induced quadrupole (inverse 8th power) term may be duplicated in this potential by a small variation in α.

4.4 THE LENNARD-JONES POTENTIAL

The general form for this potential is

$$V(r) = \lambda_n/r^n - \lambda_m/r^m \qquad (4.11)$$

The most commonly used form is however the so-called Lennard-Jones (6–12) potential, where $n = 12$ and $m = 6$. Lennard-Jones first proposed his inverse power potential to explain the experimental temperature variation of viscosity in gases on the basis that both the repulsive and attractive parts of the molecular field varied according to an inverse power of the distance.

Using the expression for the second virial coefficient of a gas in terms of $V(r)$ (see Section 4.14),

$$B(T) = 2\pi N \int_0^\infty r^2 (1 - e^{-(V(r)/kT)}) \, dr \qquad (4.12)$$

Lennard-Jones derived an expression for $B(T)$ in terms of the $(n–m)$ potential parameters. The integral contained a product of two exponentials and could be evaluated analytically as a summation containing Γ-functions. The algebra, while uncomplicated, is long and clumsy and is here omitted, as is the final expression for $B(T)$. Interested readers are referred to the original reference [5].

The parameters λ_n and λ_m for argon were originally calculated for several different values of n between 14 and 25. These were the limits resulting from an earlier attempt to explain experimental viscosity results using a repulsive force of the form $\lambda_n r^{-n}$ to represent the potential [6]. The value $m = 5$ was adopted for the attractive force index, based on the van der Waals model.

This procedure was somewhat unsatisfactory since there exist an unlimited number of potentials with n between the limits of 14 and 25, even when one assumes the reliability of these limits. For any of these potentials the λ_n and λ_m could be simply obtained by fitting to the equation of state data. Evidently more experimental criteria were required if the index n was to be considered an adjustable parameter as well as the λ_n and λ_m.

The result which permitted a limitation in n was forthcoming in the form of the experimental determination of the structure of crystalline argon. It was possible, assuming a given type of crystal structure, to sum the pair interaction energies over the crystal and thereby obtain a value for the total potential energy ϕ_0 per unit cell expressed in terms of the interatomic potential constants and the lattice parameter a. Then by minimizing this energy ($d\phi_0/da$

= 0) the resulting theoretical value of a could be compared to the experimental lattice parameter. Lennard-Jones obtained best agreement for values of n at the lower end of his previous range ($n \simeq 15$). The calculated value of a was 3.86 Å compared with an observed lattice parameter of 3.84 Å. At the same time he confirmed that with this potential the face-centered cubic (fcc) form of lattice was the most stable structure of the three possible cubic lattices.

The Lennard-Jones potential, especially in its (6–12) form, has enjoyed a high degree of popularity over the years since its introduction, although its closest equivalent, the modified Buckingham (6-exponential) potential probably has more theoretical justification than the simple power law for the repulsive term. The mixture of power law and exponential does however pose some problems in establishing parameters and in analytical treatment of atomic problems. The Lennard-Jones power potential, which resembles in general form the schematic $V(r)$ shown in Fig 1.3, was originally developed to treat inert gases, but has often been used to describe metals and other forms of solid and liquid.

4.5   THE MORSE POTENTIAL

A form of interatomic potential which dispenses completely with power law dependence was proposed by Morse [7] for the atoms of a diatomic molecule, in order to calculate its energy levels. Morse objected to the power series expansion for $V(r)$ on the grounds that when the constants were calculated from allowed energy levels of the molecule, the series did not converge for large values of r. He required a potential satisfying three conditions:

(1) $V(r) \to 0$ as $r \to \infty$;
(2) $V(r)$ has a minimum when $r = r_0$ (intermolecular separation);
(3) $V(r) \to \infty$ as $r \to 0$ (or at least becomes very large);
(4) $V(r)$ should have the same allowed energy levels as those given by the equation

$$W(n) = -D + h\omega_0\{(n + \tfrac{1}{2}) - x(n + \tfrac{1}{2})^2\} \tag{4.13}$$

Morse chose a form of potential containing two exponential terms:

$$V(r) = De^{-2\alpha(r-r_0)} - 2De^{-\alpha(r-r_0)} \tag{4.14}$$

This satisfied requirements (1) and (2), and at $r = 0$, $V(r)$ was sufficiently large to make no difference for equilibrium calculations.

A solution for the radial part of the Schrödinger equation using the above expression for $V(r)$ yielded energy levels given by

$$W(n) = -D + h\omega_0\{(n + \tfrac{1}{2}) - (h\omega_0/4D)(n + \tfrac{1}{2})^2\} \tag{4.15}$$

Thus $V(r)$ was a valid expression throughout the range of validity of this type of representation of $W(n)$. Spectroscopic data for a number of different molecules were used by Morse to calculate the constant parameters of this potential for these molecules.

It should be noted that, although the Morse form has provided quite a good representation of spectral data, the long-range attractive part of the interatomic potential is theoretically better defined by a van der Waals type of inverse power dependence than by an exponential. It was in fact shown by Linnett [8] that a better interpretation of spectroscopic constants was obtained by a mixed potential of the modified Buckingham form:

$$V(r) = Ae^{-nr} - B/r^m \tag{4.16}$$

The Morse potential is not however restricted to molecular energy level applications. Girifalco and Weizer [9] applied it to cubic metals by computing the potential parameters from various crystal properties and using them to compare calculated values of other properties with experiment. According to the Morse function, the total energy of a crystal of N atoms is

$$\Phi = D(N/2) \sum_j \{e^{-2\alpha(r_j - r_0)} - 2e^{-\alpha(r_j - r_0)}\} \tag{4.17}$$

Evaluated for a crystal with the experimental value of the lattice parameter, this energy of cohesion at equilibrium $U_s(a)$ was compared with the experimental energy of sublimation extrapolated to zero temperature and pressure. In addition, Φ should be minimum at the equilibrium separation in the crystal:

$$\Phi(a) = U_s(a) \tag{4.18}$$

$$[d\Phi/dr]_a = 0 \tag{4.19}$$

The next step was to relate Φ to the compressibility K_B:

$$1/K_B = [V \cdot d^2\Phi/dV^2]_a \tag{4.20}$$

Since $V/N = Ca^3$ where C is 4 for the bcc lattice and 2 for the fcc lattice, Eq. (4.20) could be rewritten

$$1/K_B = (1/9NCa)[d^2\Phi/dr^2]_a \tag{4.21}$$

Using Eqs. (4.18), (4.19), and (4.21) the constants of the Morse function D, α, and r_0 could be calculated. We note that r_0 is here an adjustable parameter and does not signify the nearest-neighbor separation in the crystal, as is the case for other potentials later in the present chapter (see Section 4.8). Its numerical value is, however, in most instances comparable with the nearest neighbor distance.

Girifalco and Weizer performed the calculation outlined above for a number of fcc and bcc metals, and obtained values for three parameters r_0, L, and β, where L and β were related to D and α by the equations

$$L = \tfrac{1}{2}ND, \qquad \beta = e^{\alpha r_0} \tag{4.22}$$

The results of their calculations are reproduced in Table 4.1.

TABLE 4.1[a]

MORSE POTENTIAL PARAMETERS FOR 16 METALS

Metal	αa_0	β	$L \times 10^{-22}$ (eV)	$\alpha(A^{-1})$	$r_0(A)$	D (eV)
Pb	2.921	83.02	7.073	1.1836	3.733	0.2348
Ag	2.788	71.17	10.012	1.3690	3.115	0.3323
Ni	2.500	51.78	12.667	1.4199	2.780	0.4205
Cu	2.450	49.11	10.330	1.3588	2.866	0.3429
Al	2.347	44.17	8.144	1.1646	3.253	0.2703
Ca	2.238	39.63	4.888	0.80535	4.569	0.1623
Sr	2.238	39.63	4.557	0.73776	4.988	0.1513
Mo	2.368	88.91	24.197	1.5079	2.976	0.8032
W	2.225	72.19	29.843	1.4116	3.032	0.9906
Cr	2.260	75.92	13.297	1.5721	2.754	0.4414
Fe	1.988	51.97	12.573	1.3885	2.845	0.4174
Ba	1.650	34.12	4.266	0.65698	5.373	0.1416
K	1.293	23.80	1.634	0.49767	6.369	0.05424
Na	1.267	23.28	1.908	0.58993	5.336	0.06334
Cs	1.260	23.14	1.351	0.41569	7.557	0.04485
Rb	1.206	22.15	1.399	0.42981	7.207	0.04644

[a] After Girifalco and Weizer [9].

Using a Debye model for the thermal part of the free energy in conjunction with their Morse potentials, Girifalco and Weizer obtained reasonable agreement in the calculated and experimental equation of state curves for these metals. In addition, they calculated the elastic constants and found that these agreed well with experiment in some cases though not so well in others.

From the curves of Fig. 4.2, where we compare the Morse potentials given by Table 4.1 for Cu and Na, it is evident that the interaction decays excessively slowly in the latter case. For metals such as sodium where the atoms are not tightly packed this method of obtaining constants for the Morse form overestimates the attractive part of the interaction, which depends on the screening of the ion cores by the conduction electrons (see Chapters V and VI). The range may be decreased by dividing the potential by another r-dependent exponential term and adjusting this term for a given cutoff distance. This is however a somewhat arbitrary procedure.

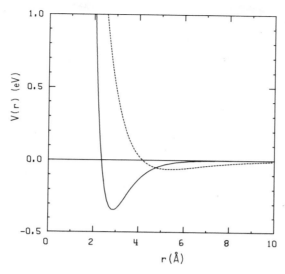

Fig. 4.2. Morse potential for Cu and Na, with parameters given by Girifalco and Weizer: (———) Cu; (- - -) Na.

4.6   ANALYTICAL POTENTIALS FOR ATOMIC COLLISIONS

With the exception of the very simplest forms such as the hard sphere interaction, the empirical potentials considered in this chapter so far have taken the form of a repulsion at small separation decreasing to an attractive minimum and finally approaching zero for large *r*. The information on which these potentials is based applies in general to very low energies, in a solid for example, where the interacting atoms are in equilibrium. Consequently, it must be expected that the potential will become more inaccurate with increasing energy and close approach of the atoms. Some other criterion must be used for potentials to be applied to energetic atomic collisions. The principal difficulty here is that experimental information on atomic interactions at high energies is very sparse, the only reliable source being the results of atom–atom scattering experiments, which are difficult to perform in a manner which yields sufficiently accurate data (see Chapter VIII).

For this reason interatomic potential parameters for higher energy collisions are based on purely theoretical considerations to a higher degree than those which apply to equilibrium situations. Their accuracy increases with increasing energy (provided that they have a built-in nuclear repulsion) since at high energy and small separation the approximations for the many-body effects of the atomic electrons have greater validity. There remains unfortunately an intermediate range of energy in which the many-body nature of the interaction is important and the potential is not well defined. This region is particularly important in the field of radiation damage, which would benefit greatly from additional experimental results in atom–atom scattering or other nonequilibrium phenomena (see Sections 4.15 and 4.16).

In the absence of an *a priori* knowledge of the form of the potential in the intermediate range, it has been necessary to extrapolate to this region potentials which are valid at either high energies or near equilibrium. In the next few sections we shall discuss some forms of potential applicable to atomic collisions and some methods of performing the extrapolation to the intermediate energy range. We note that for interaction energies greater than a few electron volts, the attractive part of the potential is of no consequence and in most models that are used for atomic collisions it is neglected.

4.7 SCREENED COULOMB POTENTIALS

We have already considered at length one potential form applicable to high energy interactions, namely the TF potential (see Chapter III). This and other similar forms are all based on a limiting simple Coulomb repulsion between the nuclei at very small values of r, combined with a form of screening function to simulate the interaction between the atomic electrons, which makes itself felt at large separations:

$$V(r) = (Z_1 Z_2 e^2 / r) f(r) \tag{4.23}$$

where $f(r)$ is the screening function, having the boundary conditions

$$f(0) = 1 \tag{4.24}$$

$$f(\infty) = f'(\infty) = 0 \tag{4.25}$$

The simplest screened Coulomb potential was suggested by Bohr [10] and was exponential in the form

$$f(r) = e^{-r/a} \tag{4.26}$$

where the screening radius

$$a = a_0 (Z_1^{2/3} + Z_2^{2/3})^{-1/2} \tag{4.27}$$

This potential decays very rapidly with increasing r, becoming invalid at separations greater than a few tenths of an angstrom. Its validity may be extended slightly by treating the screening radius as a variable parameter, although it is still applicable only to relatively high energy collisions ($\gtrsim 100$ keV).

Brinkman [11] suggested a modification of the exponentially screened Coulomb potential, which is arrived at by calculating the interaction energy of two nuclei each possessing a rigid charge distribution and a screened Coulomb field of the simple Bohr type. This was of the form

$$f(r) = \frac{a_1^2 e^{-r/a_2} - a_2^2 e^{-r/a_1}}{a_1^2 - a_2^2} \tag{4.28}$$

where a_1 and a_2 are the screening radii of the two atoms. For identical atoms upon expansion, the potential becomes

$$f(r) = (1 - r/2a)e^{-r/a} \tag{4.29}$$

where $a_1 = a_2 = a$. This potential has become known as the Brinkman 1 potential. When r is small compared to the screening radius the interaction becomes exponentially screened Coulomb in nature. It becomes attractive for r greater than $2a$. We note, however, that the physical basis of a rigid charge distribution remaining unaltered during a collision, coupled with the neglect of exchange interactions and a closed shell repulsion, is no more correct than that leading to the simpler form of Eq. (4.26). In fact the Brinkman 1 potential decreases even more rapidly with increasing r than the Bohr potential (see Fig. 4.3). The latter is therefore considered preferable in cases where an exponentially screened Coulomb potential is applicable.

The other common type of screened two-body interaction is that obtained by Firsov based on TF theory. The screening function is that found by the solution of the TF dimensionless equation (2.23):

$$f(r) = \chi(r/a) \tag{4.30}$$

where now

$$a = 0.885a_0(Z_1^{1/2} + Z_2^{1/2})^{-2/3} \tag{4.31}$$

Although the screening function is not analytical and consequently falls outside the scope of this chapter, it is accurately known (see Appendix 1) and the Firsov potential is normally included in the list of empirical potentials.

While this potential resembles the exponentially screened Coulomb form at small separations, the infinite electron distribution of the TF atom drives it to the opposite extreme, causing it to decay much too slowly with increasing distance. The inclusion of exchange in a TFD treatment removes this difficulty but introduces others connected with the discontinuous electron distribution which have not yet been completely overcome (see Chapter III).

A comparison of the above types of screened Coulomb potential for Cu is illustrated by Fig. 4.3. None of the three forms is satisfactory at large separations, the Bohr and Brinkman 1 potentials decaying too rapidly and the Firsov too slowly as r increases. Their application is limited to relatively high energy interactions between atoms in which the nuclei approach each other to a minimum separation of less than a few tenths of an angstrom.

Instead of using the accurate tabulated values of the TF function $\chi(r/a)$ in Eq. (4.30), the screening function could be any of the analytical approximations listed in Section 2.8. Some of these may even represent more realistically

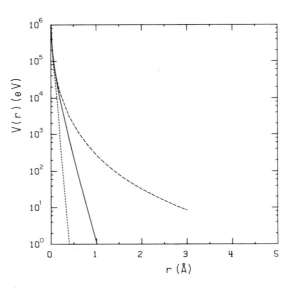

Fig. 4.3. Comparison of screened Coulomb potentials for Cu: (——) Bohr; (---) Brinkman; (–––) Firsov.

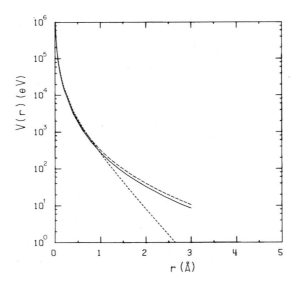

Fig. 4.4. Comparison of TF potentials for Cu: (——) Firsov; (---) Molière; (–––) Sommerfeld.

the interatomic potential in the intermediate r range (see Section 4.11). Figure 4.4 compares the interatomic potentials for Cu resulting from analytical approximations of the Molière and Sommerfeld type (see Eqs. (2.46) and (2.63)) with the Firsov form using the tabulated TF function.

4.8 THE BORN–MAYER POTENTIAL

This simple two-parameter exponential potential, of the form

$$V(r) = Ae^{-Br} \tag{4.32}$$

was introduced by Born and Mayer [12] in their treatment of ionic crystals, to represent the closed shell repulsion of the ions. Although simple it is theoretically sound since an exponential form is suggested by quantum mechanical considerations.

From the finite behavior of this potential, as r tends to zero we see that its validity is restricted to separations involving only moderate interpenetration of the closed electron shells. The deeper the penetration the less accurate is the Born–Mayer potential, since in particular no nuclear repulsion enters the interaction. In fact, if as is usually the case the parameters A and B are determined from equilibrium crystal data, the Born–Mayer (BM) potential is strictly speaking valid only for very small overlap of the closed shells. It is however reasonable to apply it to atomic interaction energies up to some-where in the neighborhood of one half of the parameter A.

Huntington [13] revised the BM form slightly for application to crystal lattices, introducing the equilibrium nearest-neighbor separation r_0 into the equation

$$V(r) = A' \exp\{-\rho(r - r_0)/r_0\} \tag{4.33}$$

There are various methods of establishing values for the constant parameters of the BM potential, using data taken from both equilibrium and higher energy experimental situations. In the next section we shall consider the adjustment of the parameters to crystal elastic constant data. Later in the present chapter and in Chapter VII the adjustment of this and other simple potentials to the results of experiments involving dynamic collision processes will be discussed.

4.9 ADJUSTMENT OF THE BORN–MAYER POTENTIAL TO CRYSTAL ELASTIC CONSTANTS

When adjusting an analytical pair potential for metals to elastic constant data it must be remembered that it is not only the ion–ion closed shell repulsion which contributes to the experimental elasticity. The effects of the conduction electron gas must also be taken into account. To a reasonable approximation the ion–ion and electron parts of the elastic moduli may be distinguished and calculated separately, under the assumption of free or nearly free electrons. A method of estimating the separate contributions has been described by Fuchs [14]. We shall not go into the details of the calculation here, but simply indicate the physical principles and the most important equations. A good review of elastic constants in crystals is given by Huntington [15].

Consider the case of a fcc crystal. We use the normal notation for the crystal elastic constants: thus the bulk modulus K_B is $(c_{11} + 2c_{12})/3$, and the two shear moduli are $(c_{11} - c_{12})$ and c_{44}. If a crystal can be described satisfactorily using only central forces between atoms, it can be shown that the experimental elastic constants should satisfy certain rules known as the Cauchy relations. There is only one Cauchy relation for a cubic crystal, namely $c_{12} = c_{44}$, and this is not satisfied for any metallic crystal. The experimental constants contain contributions from the ion–ion repulsion and from the electron gas, which has itself a certain compressibility, although it contributes little to the shear moduli. We denote by primes on the elastic constants the parts which result from the short-range forces represented by the two-body potential $V(r)$. Then the primed elastic constants are given in terms of the potential by the equations:

$$\frac{1}{3}(c'_{11} + 2c'_{12}) = \frac{2}{3} N r_0{}^2 \left[\frac{d^2 V}{dr^2}\right]_{r_0}$$

$$c'_{44} = \frac{1}{2} N \left\{ r_0{}^2 \left[\frac{d^2 V}{dr^2}\right]_{r_0} + 3 r_0 \left[\frac{dV}{dr}\right]_{r_0} \right\} \qquad (4.34)$$

$$c'_{11} - c'_{12} = \frac{1}{2} N \left\{ r_0{}^2 \left[\frac{d^2 V}{dr^2}\right]_{r_0} + 7 r_0 \left[\frac{dV}{dr}\right]_{r_0} \right\}$$

Here r_0 is the nearest-neighbor separation and N the number of ions per unit volume. Only nearest-neighbor interactions go to make up these equations. They may be solved for c'_{11}, c'_{12}, and c'_{44}, eliminating N using the fact that for a face-centered cubic lattice

$$N = 4/(\sqrt{2}\,r_0)^3 \qquad (4.35)$$

The primed elastic constants are given by the equations

$$c'_{11} = \frac{\sqrt{2}}{r_0} \left\{ \left[\frac{d^2 V}{dr^2}\right]_{r_0} + \frac{1}{r_0} \left[\frac{dV}{dr}\right]_{r_0} \right\}$$

$$c'_{12} = \frac{1}{\sqrt{2}\,r_0} \left\{ \left[\frac{d^2 V}{dr^2}\right]_{r_0} - \frac{5}{r_0} \left[\frac{dV}{dr}\right]_{r_0} \right\} \qquad (4.36)$$

$$c'_{44} = \frac{1}{\sqrt{2}\,r_0} \left\{ \left[\frac{d^2 V}{dr^2}\right]_{r_0} + \frac{3}{r_0} \left[\frac{dV}{dr}\right]_{r_0} \right\}$$

Without a reliable estimate of the electronic contribution to the elastic moduli it is difficult to determine the potential constants uniquely. The bulk modulus contains a term arising from the pressure of the electron gas, which in turn depends on the degree of freedom permitted to the electrons. If a free-electron model is assumed, this contribution to K_B for Cu is 6.4×10^{11} dyn/cm^2. If on the other hand, an effective mass taken from specific heat measurements is used to calculate the electron gas compressibility, the corresponding contribution is 4.3×10^{11} dyn/cm^2 [13]. Fuchs [14] calculated the influence of the long-range electrostatic interaction of the ions with the conduction electrons on the shear moduli, obtaining electronic contributions of 0.57×10^{11} dyn/cm^2 to $(c_{11} - c_{12})$ and 2.57×10^{11} dyn/cm^2 to c_{44}. The comparative values of these two terms have an effect on the anisotropy, which depends on $2c_{44}/(c_{11} - c_{12})$, and it turns out that the BM parameter B is sensitive to this effect. In a variational calculation of what fraction of the Fuchs electronic contribution should be permitted to influence the shear distortion in an fcc crystal, Huntington [13] estimated that the Fuchs terms should be reduced by at least 25%, and that a more elaborate calculation would reduce them by a half.

Making use of the set of electronic contributions including the nearly free electron gas compressibility, we might calculate the constants for a BM type of potential such as that given by Eq. (4.32) or Eq. (4.33). Table 4.2 gives the experimental data and resulting primed elastic moduli. From this table we have $c'_{11} = 13.6$, $c'_{12} = 8.1$, and $c'_{44} = 5.6$ in units of 10^{11} dyn/cm^2. Substituting these values into Eq. (4.36) yields figures for the BM parameters A' and ρ of Eq. (4.33) or A and B of Eq. (4.32). We note, however, that since there are two parameters and three equations the two constants will not be uniquely defined (we need not expect consistency owing to the approximative nature of

TABLE 4.2

ELECTRON GAS AND ION REPULSION CONTRIBUTIONS TO
EXPERIMENTAL ELASTIC MODULI OF COPPER

	$(c_{11} + 2c_{12})/3$[a]	c_{44}[a]	$c_{11} - c_{12}$[a]
Experiment[b]	14.9	8.2	5.1
Electronic contributions	4.3	2.6	0.6
Primed moduli	10.6	5.6	4.5

[a] Units in 10^{11} dyn/cm^2.
[b] Overton and Gaffney [16].

the estimation of the electronic effects). One method of resolving the imprecision is to neglect the compressibility as being the least reliable from the electron point of view, and to solve Eq. (4.34) for c'_{44} and $(c'_{11} - c'_{12})$. Even so, the uncertainty inherent in the electron terms for the shear constants leads to a considerable spread in the parameters depending on the values assumed. This point is illustrated by Table 4.3, in which the BM parameters are calculated assuming 0, 50, and 100% of Fuchs's values. The resulting three potentials for Cu are plotted in Fig. 4.5.

TABLE 4.3

BM Potential Parameters for Cu Assuming Different Electron Gas Contributions to Experimental Elastic Moduli

Electron term assumption	A' (eV)	ρ (Å)	A (eV)	B (Å^{-1})
Fuchs's values	0.0178	23.0	1.74×10^8	9.03
50% of Fuchs's values	0.0479	16.1	4.58×10^5	6.32
No electron term	0.0631	13.5	4.57×10^4	5.32

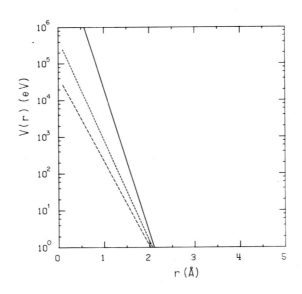

Fig. 4.5. BM potential for Cu using different fractions F of the Fuchs electronic contributions: (——) $F = 0$; (- - -) $F = 0.5$; (— — —)$F = 1.0$.

An alternative method of defining the two parameters from the three elastic equations is the least squares procedure used by Huntington [13] to obtain A' and ρ with a deviation of less than 6%. His first potential included an electronic contribution equivalent to 50% of the Fuchs term. In the same publication he proposed two other potentials, the first taking into account the variation of compressibility with pressure (which gives a slightly lower ρ) and the second with a larger ρ such as might be found by incorporating the full Fuchs electronic terms. The constants A' and ρ for the three Huntington Cu potentials are given in Table 4.4, together with the equivalent A and B. It should be noted that the experimental elastic constants used by Huntington varied slightly from those quoted in Table 4.2.

TABLE 4.4

PARAMETERS FOR 3 BM POTENTIALS FOR Cu[a]

Huntington potential	A' (eV)	ρ (Å)	A (eV)	B (Å^{-1})
1	0.056	14.9	1.657×10^5	5.84
2	0.053	13.9	5.769×10^4	5.45
3	0.038	17.2	1.121×10^6	6.74

[a] Suggested by Huntington [13].

The discussion above is by no means a comprehensive review of the determination of potentials from crystal elasticity data. In particular, force constants and interatomic potentials which include electron gas effects can be obtained either directly or indirectly from experimental phonon spectra (see Section 6.9). In this section we have limited ourselves to the closed shell repulsion between ions, as represented by the BM potential, and pointed out the uncertainties due to the presence of the electron gas. The conclusion is that the availability of experimental elastic data in the absence of more precise information on electronic effects leaves us unable to determine uniquely the BM potential. Some further knowledge is necessary to reduce the amount of choice available. A good example of how to select a potential from those suggested by elastic data was provided by the work of Gibson et al. [17] (GGMV), who attempted to find a potential suitable for their atomistic computer simulations.

The approach of GGMV was based on the fact that the possible range of BM interactions varies by only a small amount close to the equilibrium separation, but since the logarithmic slope depends on ρ the divergence of the different potentials increases with decreasing r (see Fig. 4.5). They chose their potential from three possible ones, the constants of which are given in Table 4.5. Their method of selecting one interaction from these three was to find the one which most closely coincided with the TF form for copper and the TFD form as calculated by Abrahamson (see Chapter III), for the region of atomic separation between 0.5 and 1.0 Å, corresponding to energies between about 200 and 2000 eV. In fact, their potential 2 was consistent with both TF and TFD forms at these values of r, although it agreed better with the latter (Fig. 4.6). GGMV therefore chose potential 2 for their computation work, and the so-called "Gibson 2" potential has since been extensively used as a pair interaction for Cu.

TABLE 4.5

PARAMETERS OF 3 BM POTENTIALS FOR Cu[a]

GGMV potential	A' (eV)	ρ (Å)	A (eV)	B (Å^{-1})
1	0.0392	16.97	9.189×10^5	6.65
2	0.0510	13.00	2.257×10^4	5.09
3	0.1004	10.34	3.102×10^3	4.05

[a] Suggested by Gibson et al. [17].

Following the work of Gunther (see section 3.5), which pointed out errors in the Abrahamson TFD potential, one might ask whether the Gibson 2 potential remains valid. In fact the true TFD potential must lie below the TF potential over the whole range of r. It is evident from Fig. 4.6 that potential 1 represents an excessively hard atom. Possibly a true TFD potential might almost coincide with potential 3, but one must take into account the fact that ρ for potential 3 is outside the range suggested by the elastic constant data (Table 4.3) and represents an extremely soft atom. Therefore, since potential 2 agrees well with the TF potential it remains the best choice of the three. GGMV were also aided in their choice by the fact that in the computer simulation the potential 2 gave an atomic displacement threshold energy in closest agreement with experiment [17].

This idea, touched upon by GGMV, of combining two or more analytical forms of potential valid over different regions of separation, has received widespread attention. It represents a good method of obtaining an interaction valid over the difficult intermediate range where screened Coulomb potentials fail and equilibrium data does not apply. There are two basic methods of approach. One is to incorporate the two forms in one composite analytical

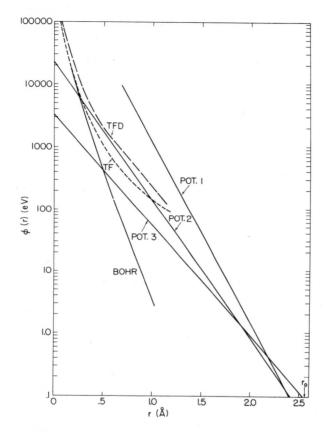

Fig. 4.6. Comparison of GGMV potentials 1, 2 and 3 from Firsov (TF) and Abrahamson (TFD) potentials; r_0 is the nearest-neighbor distance. (After Gibson *et al.* [17]).

formula, while the other is to allow each its own range of r and adjust parameters to ensure analytic continuation over the boundary. We shall consider some such mixed potentials in the next section.

4.10   MIXED ANALYTICAL FORMS OF POTENTIAL

Of the analytical potential forms which we have considered it seems that a screened Coulomb interaction applies best to small separations, and a BM, Morse, or similar form to equilibrium separation of greater. It is therefore reasonable to suggest that if two or more types were combined the intermediate range of r might be better represented by the resulting potential than by any one separately. One such composite potential was proposed by Brinkman [18] as a compromise between the exponentially screened Coulomb and the BM types:

$$V(r) = AZ_1Z_2e^2 \cdot \frac{e^{-Br}}{1 - e^{-Ar}} \tag{4.37}$$

It may be easily seen that this reduces to the two simple forms at the limits of r:

$$\begin{aligned} V(r \to 0) &= (Z_1Z_2e^2/r)e^{-Br} \\ V(r \to \infty) &= AZ_1Z_2e^2\,e^{-Br} \end{aligned} \tag{4.38}$$

Thus this "Brinkman 2" potential approximates a screened Coulomb potential at small r and a BM at large r. The constant B in Brinkman's treatment was given by

$$B = (Z_1Z_2)^{1/6}/Ca_0 \tag{4.39}$$

where C is an adjustable constant near unity. Determination of A and B from elastic constant data showed that C could be made equal to 1 for atomic number $Z \leq 30$, but that for heavy solids, to which Brinkman wished to apply his model, it was necessary to make C equal to 1.5. For Cu, Ag, and Au he gave an empirical value of A, calculated from elastic constants:

$$A = (0.95 \times 10^{-6}/a_0)Z^{7/2} \tag{4.40}$$

The resulting potential would be best applied to heavy solids.

Another such composite potential was developed by Leibfried [19] for the particular case of Cu, again relying on the exponentially screened Coulomb potential at small separations and the BM form at large distances. It was of the form

$$V(r) = E_a(\tfrac{1}{4} + a/r)e^{-r/a} \tag{4.41}$$

Chosen to reproduce the experimental elastic data, the constants were

$$\begin{aligned} E_a &= 7.5 \times 10^4 \quad \text{eV} \\ a &= r_0/13 \simeq 2 \times 10^{-9} \quad \text{cm} \end{aligned} \tag{4.42}$$

The screened Coulomb and BM forms are especially suited to analytical combination since both limiting potentials contain an exponential term (although not necessarily the same or even compatible for all elements). Another method of combination consists not of setting up a composite potential valid over the whole range, but of fixing limiting values of r for each simple form and adjusting the potential parameters so that the separate forms are analytically continuous where they meet. It then becomes a question of deciding the exact location of the boundary, which is not simple since neither high-energy nor equilibrium potentials are very accurate in the intermediate region. Indeed, it is often largely a matter of personal choice, which makes it important in any calculation using such a theoretical potential to investigate the sensitivity of the results to this pair interaction.

One general method of obtaining a composite potential reasonably valid over a large separation or energy range has been presented by Genthon [20]. Essentially, this potential consists of a TF Firsov potential $V_{TF}(r)$ at short distance and a BM form at the equilibrium separation $V_{BM}(r)$. The two forms are constrained by analytical continuity at a variable changeover point r_c and by compatibility of the BM potential with the bulk modulus K_B, after the free electron and band-structure energy contributions have been removed from the latter (see Section 4.9). The use of the bulk modulus is feasible only in close-packed metals such as Cu, Ag, and Au since in the alkalis and polyvalent metals the electron contribution to the measured compressibility is predominant and hence K_B is small and ill-defined. As we have already noted, even in the case of the noble metals there remains a significant degree of uncertainty in defining the role of the conduction electrons.

For the TF potential Genthon made use of the Latter approximation to the TF screening function (see Section 2.7) and the Firsov form of screening radius as defined by Eq. (4.31). This uniquely defines the potential for any atomic number at small separations.

It remains to determine the two parameters of the BM potential and the changeover point r_c—three constants in all. To achieve this there are three conditions, the first represented by the compressibility part of Eq. (4.34) where $V(r) = V_{BM}(r)$. The other two conditions rest in the continuity of the potential at $r = r_c$:

$$V_{TF}(r_c) = V_{BM}(r_c)$$
$$[dV_{TF}/dr]_{r_c} = [dV_{BM}/dr]_{r_c} \tag{4.43}$$

In a least-squares analysis for Z ranging from 12 to 92, Genthon found that it was a reasonable approximation to assume that the atomic number did not influence to any significant extent the BM exponential constant B, which could be related to the nearest-neighbor separation r_0 by the equation

$$B = 6.11 r_0^{-0.25} \tag{4.44}$$

The values of B obtained by Genthon are given in Table 4.6 for a number of metals. The small variation in B suggests that a further simplification might be to consider it as invariant with respect to Z. However, this assumption would cause rather too great a dispersion in calculated values of K_B, which is sensitive to this constant.

TABLE 4.6^a

Parameters for the Genthon Mixed Analytical Potential for 6 Metals

Metal:	C	Mg	Cu	Ag	Au	U
B (Å^{-1})	4.95	4.57	4.95	4.57	4.65	4.61
ξ	0.925	1.011	1.019	0.994	0.948	0.923
A (eV)	2,088	5,381	25,342	42,691	94,388	113,511
r_i (Å)	0.41	0.50	0.51	0.56	0.58	0.61

^{a} The values of B for Cu, Ag, and Au were determined directly from the compressibility, while those for C, Mg, and U were calculated using Eq. (4.44). The use of this equation for the noble metals would change the values of B by less than 3%. We note that once r_i and B are known, A may be obtained directly from the formula $A = V_{TF}(r_i)e^{Br_i}$.

The constant A of the BM form is given in Genthon's formulation in terms of B and the TF potential constants, again by a curve-fitting procedure, by the equation

$$A = \xi(Z^2e^2/\mu)(\mu B)^{2.4} \qquad (4.45)$$

where

$$\mu = 0.885a_0/\sqrt{2}\,Z^{1/3} \qquad (4.46)$$

and ξ is a constant which is ~ 1 for all but the very light or very heavy elements. Both ξ and A are included for the six elements of Table 4.6.

We compare in Fig. 4.7 the three composite potentials for Cu which we have described above, namely the Brinkman 2, Leibfried, and Genthon forms. In the case of the Brinkman 2 potential we have assumed that the value of the constant C is 1. Of the three, the Genthon potential is perhaps the best, being at the same time the most adaptable and the most soundly based on experimental data.

4.11   The Molière Potential

One simple interatomic potential which, although not a mixed form fulfills the task of the latter, is that obtained from the Molière approximation to the TF screening function (see Eq. (2.63) and Fig. 4.4). Paradoxically, since the Molière exponential form of $\chi(x)$ diverges from the accurate TF $\chi(x)$ at large values of x (after about 6 screening lengths), it leads to a more accurate form of interatomic potential given by the equation

$$V(r) = (Z_1Z_2e^2/r)\{0.35e^{-0.3r/a} + 0.55e^{-1.2r/a} + 0.10e^{-6.0r/a}\} \quad (4.47)$$

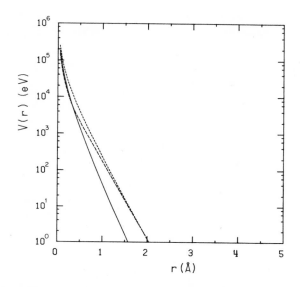

Fig. 4.7. Comparison of some mixed analytical potentials for Cu: (——) Brinkman [18]; (- - -) Leibfried [19]; (– – –) Genthon [20].

where a is the Firsov screening length of Eq. (4.31). The point is illustrated very clearly in a comparison by Robinson [21] of the screening functions involved in TF, Bohr, and BM Cu–Cu and Au–Au potentials, reproduced in Fig. 4.8. The screening function $f(r)$ is given by

$$f(r) = rV(r)/Z_1 Z_2 e^2 \qquad (4.48)$$

One can see quite clearly that while the TF function falls off too slowly at large r, the Molière approximation decreases more rapidly, and is in good agreement with the BM forms for Cu and Au in the region of the equilibrium separation for these metals. The Au–Au BM potential used here is one suggested by Thompson [22], where the constants are based on compressibility and focusing energy in crystals:

$$V(r) = 2 \times 10^5 e^{-5.0r} \qquad [r \text{ in Å}, \quad V(r) \text{ in eV}] \qquad (4.49)$$

Since the Molière screening function is only an approximation to the TF screening function there is not necessarily any justification for using the Firsov form of screening radius in the Molière interatomic potential. It can in fact be treated as an adjustable parameter. This has been done in the Molière potentials for Cu–Cu, Fe–Fe and Au–Au interactions which are compared with Firsov and BM forms in Fig. 4.9. The curves fall between the Bohr and Firsov potentials for r greater than ~ 1 Å and should therefore be better

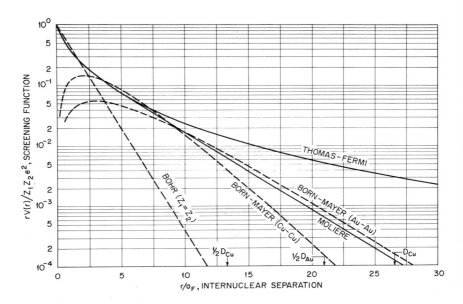

Fig. 4.8. Screening function of the Molière potential compared with those of Firsov and Bohr and related functions for Cu–Cu and Au–Au BM potentials. The unit of distance is the Firsov screening length a_F; D represents the nearest-neighbor distance in the lattice. (After Robinson [21].)

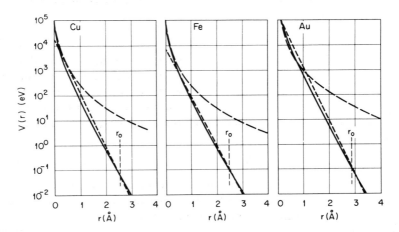

Fig. 4.9. Molière potentials for Cu, Fe, and Au compared to Firsov and BM forms. The Molière screening distance has been adjusted so that the potential coincides with the BM potential at the nearest-neighbor separation in the crystal (r_0): (–––) Firsov; (-·-) Born–Mayer; (——) modified Molière.

approximations than either of these two for low and intermediate energy atomic collisions occurring in for example radiation damage studies.

4.12   COMPOSITE POLYNOMIAL POTENTIALS

A different method of setting up a mixed or composite potential is illustrated by the work of Johnson [23], who developed an empirical short-range ionic interaction potential extending to between second and third neighbor distances in bcc crystals. A cutoff after the first-neighbor distance would be invalid since the bcc structure is unstable when only first-neighbor interactions are taken into account. This form of potential was developed to match the short-range elastic moduli, again including second neighbor interactions.

If we assume the first and second derivatives of the potential to be determined at first- and second-neighbor distances r_1 and r_2, and dV/dr to be zero at the potential cutoff r_c, then the simplest form of dV/dr which satisfies these conditions is a parabola whose coefficients may be fitted to the elastic moduli. Unfortunately this parabola does not go to zero at $r = r_c$ since the cutoff does not enter the elastic moduli equations. It is therefore necessary to match another parabola to the first at a point just greater than $r = r_2$, forming a smooth curve which is tangential to zero at the cutoff distance. The actual potential $V(r)$ is then found by integration, again using the cutoff condition to determine the constant of integration. Finally, for small values of r below r_1 the now cubic potential curve is matched to a third cubic potential which is itself matched to the dynamic radiation damage potential of GGMV for alpha-iron at a distance well below r_1. Thus Johnson's interaction potential was a composite of three cubic equations, each of the form

$$V(r) = A(r - B)^3 + Cr + D \qquad (4.50)$$

The values of the parameters A, B, C, and D found by Johnson over the different ranges of separation for Fe and V are reproduced in Table 4.7 [24], and the resulting curves shown in Fig. 4.10.

Johnson used this interatomic potential in a computer simulation of interstitials and vacancies in various configurations in a crystal lattice of alpha-iron. The purpose was to estimate the stability of various point defect configurations and obtain values for the migration energies of these defects, for comparison with experimental annealing studies. While this is a good method of studying the point defect energy problem from a theoretical point of view, it is rather easy to attack these computations in their most vulnerable flank, namely the choice of interatomic potential. In later work Johnson showed that some of the numerical values of defect energies were rather sensitive to the point at which his pair interaction was cut off [25]. He

TABLE 4.7

PARAMETERS FOR THE JOHNSON POLYNOMIAL POTENTIAL FOR Fe AND V

Metal	r (Å)	A	B	C	D
Fe	1.90 — 2.40	−2.195 976	3.097 910	2.704 060	−7.436 448
	2.40 — 3.00	−0.639 230	3.115 829	0.477 871	−1.581 570
	3.00 — 3.44	−1.115 035	3.066 403	0.466 892	−1.547 967
V	2.00 — 2.53	−1.496 112	2.731 297	−0.599 656	1.614 190
	2.53 — 3.17	0.095 514	0.304 042	−2.201 314	4.625 125
	3.17 — 3.63	−1.430 039	3.361 421	0.309 466	−1.095 655

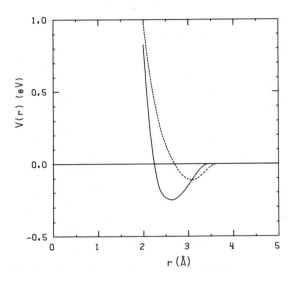

Fig. 4.10. Johnson composite polynomial potential for Fe and V: (——) Fe; (---) V.

suggested that in fact it might be desirable to adjust the value of r_c until the formation energy of a vacancy coincides with that found experimentally.

One specific criticism of the Johnson type of potential is that its approach to zero is artificial in that it is unrealistically rapid. It is possible that Johnson's results might be less potential-sensitive if his computations were carried out with a potential which did not disappear completely but was rather cut off at a small negative value. This would avoid the extremely rapid variation of the interatomic force $(-dV/dr)$ close to the zero potential, which could significantly affect the relaxation of the lattice around the defect.

The polynomial technique has recently been refined and extended by Englert *et al.* [26] to obtain interatomic potentials for Fe and Cu. Their potential was made up of a volume-dependent term arising from the electron gas in addition to the pair interaction term. The pair function was constructed from a spline function containing 10 interpolated cubic polynomials, of the form

$$V(r) = A_k(r - r_k)^3 + B_k(r - r_k)^2 + C_k(r - r_k) + D_k \qquad (4.51)$$

where k varies between 1 and 10 depending on the region of the potential curve under consideration. The parameters A_k, B_k, C_k, and D_k were determined by adjusting to a number of conditions. First, $V(r)$ was matched to the GGMV potential 2 BM repulsion for distances less than the nearest-neighbor equilibrium separation. It was then required that the pair plus volume-dependent terms of the potential should lead to the correct type of crystal lattice and experimental lattice spacing at equilibrium. Third, the pair potential derivatives should correctly reproduce first- and second-neighbor force constants computed directly from phonon dispersion data. Thus the final form of the potential should include implicitly the free electron effects. As a fourth condition, the potential when cut off after the second-neighbor distance should compute correctly the experimental vacancy formation energy and stacking-fault energy, allowing some 10% for relaxation of the lattice.

The resulting pair potential for copper contained a minimum of about -0.3 eV close to the nearest-neighbor distance and a small secondary maximum at about the second-neighbor separation. The form of the potential is illustrated in Fig. 4.11, and the values of the parameters of the spline function

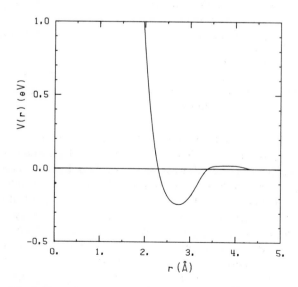

Fig. 4.11. Bullough composite polynomial potential for Cu.

are reproduced in Table 4.8. For iron the corresponding potential was slightly more complicated in form, containing two minima and corresponding maxima all below the zero in energy [27].

TABLE 4.8

PARAMETERS FOR THE POLYNOMIAL POTENTIAL OF EQ. (4.51) FOR COPPER

k	r_k (Å)	A_k	B_k	C_k	D_k
1	1.0	−667.945 8	1081.883 8	−628.564 9	138.110 0
2	1.5	−49.044 9	79.965 5	−47.640 8	10.805 0
3	2.0	−3.238 2	6.398 1	−4.459 1	0.845 3
4	2.551	−0.258 148	1.045 285	−0.357 854	−0.210 930
5	3.061 199	−2.221 407	0.650 164	0.507 164	−0.155 699
6	3.341 810	1.507 669	−1.219 882	0.347 295	−0.011 272
7	3.607 658	−0.080 144	−0.017 445	0.018 353	0.023 168
8	4.209 149	2.186 182	−0.162 063	−0.089 620	0.010 455
9	4.311 190	−1.575 972	0.507 171	−0.054 405	0.001 945
10	4.418 461	0.0	0.0	0.0	0.0

The polynomial representation, containing a large number of adjustable constants, is probably the best method of obtaining a potential which reproduces a large volume of experimental results. The final form is much too cumbersome for use in analytical theory, and is specifically designed for computer solutions of atomic problems. The remainder of the present chapter will be devoted to the adjustment of simpler potential forms to various types of experimental phenomenon.

4.13   DETERMINATION OF INTERATOMIC POTENTIAL PARAMETERS FROM EXPERIMENT

On several occasions during the preceding sections we have described the use of experimental data to define the parameters of an analytical form of potential. In particular, the adjustment of the BM form to elastic data in crystals was described in detail in Section 4.9. Let us now consider in general the conditions which apply to this process of adjustment. First, it is necessary to find a suitable experimental situation in which the data may be expressed theoretically in terms of the pair potential parameters. Then the atomic problem to which the resulting potential is to be applied must not be so far removed from the experimental conditions leading to the parameter determination that these conditions no longer apply. For example, potentials obtained from equilibrium data in a solid must be applied with caution to higher energy interactions. In addition, sufficient experimental data must exist for a unique parameter determination.

Adjustment of simple forms of potential has been carried out mainly in connection with experiments in an equilibrium or near-equilibrium situation of an assembly of atoms, whether solid, liquid or gas. Examples of the data used are elastic constants and phonon dispersion spectra in crystals, liquid structure factors, and virial coefficients in gases. Some empirical potentials have been obtained at higher energies from experiments on the scattering of atoms or ions on each other. The use of elastic data in crystals has been described Section 4.9. The subject of phonon spectra will be described in Section 6.9, when we discuss the application of pseudopotential-derived pair potentials. Determination of potentials from experiments on liquid structure is a rapidly expanding topic and merits a chapter of its own (Chapter IX). Similarly, potentials for atom–atom scattering are treated in some detail in Chapter VII. It remains to discuss the relation between virial coefficients in gases and the interatomic potential, and the determination of $V(r)$ from radiation damage and shock wave data in solids.

4.14   THE VIRIAL COEFFICIENTS

Some information on interatomic forces may be derived from the equation of state of a gas since the pressure and volume variation with temperature evidently depend on the interaction of the atoms. The perfect gas equation is of course inaccurate, and the van der Waals modified PVT equation is of limited usefulness since it is not possible to relate its constants to the forces between atoms or molecules. However, the experimental data may be fitted over a large range of temperature and pressure by the *virial equation*

$$\frac{Pv}{RT} = 1 + \frac{B(T)}{v} + \frac{C(T)}{v^2} + \frac{D(T)}{v^3} + \cdots. \tag{4.52}$$

The $B(T)$, $C(T)$ etc. are the *virial coefficients* of second, third, and higher order.

Using statistical mechanics it is possible to express the virial coefficients in terms of the interatomic potential, and consequently to obtain information about atomic interactions by analysis of PVT data in the gaseous state. Essentially, the nth virial coefficient expresses the deviation from perfect gas behavior due to n-body collisions in the gas. For a simple two-body potential we are therefore interested in the second virial coefficient $B(T)$ and would like to be able to neglect the higher terms in the virial equation. The convergence of this equation becomes gradually less satisfactory with increasing gas density until it finally diverges at about the density of the liquid. Nevertheless, the second coefficient may be adequate at low densities to explain the deviation from the ideal gas.

The equation of state may be derived from the statistical thermodynamical equation relating the pressure to the partition function Z for the system

$$P = kT[\partial(\ln Z)/\partial v]_T \tag{4.53}$$

Here Z for the nonideal gas contains the interatomic potential through the Boltzmann factor $e^{-V(r)/kT}$. The potential may or may not be assumed to be two-body, although this assumption is generally made even when we wish to calculate higher-order virial coefficients in order to simplify the mathematics. The development of the above equation using the methods of statistical mechanics leads to the following expression for the second virial coefficient:

$$B(T) = 2\pi N \int_0^\infty r^2 \{1 - e^{-V(r)/kT}\} \, dr \tag{4.54}$$

If the empirical expression for $V(r)$ is sufficiently simple it may be possible to perform the integration analytically, otherwise it is necessary to find the potential parameters by a numerical adjustment to the values of $B(T)$ at different temperatures. A simple example is that of the square-well potential of Eq. (4.2). Integration of Eq. (4.54) for this potential leads to the following equation for $B(T)$:

$$B(T) = \tfrac{2}{3}\pi N\sigma^3 \{1 - (R^3 - 1)(e^{\varepsilon/kT} - 1)\} \tag{4.55}$$

with the quantities σ, ε, and R defined as in Eq. (4.2). More complicated expressions result from the integration of other analytical interactions such as the Lennard-Jones potential.

The method normally used to deduce the second virial coefficient from equation of state data is to limit the virial equation to a finite polynomial instead of an infinite series:

$$\frac{Pv}{RT} = 1 + \frac{A_1(T)}{v} + \frac{A_2(T)}{v^2} + \cdots + \frac{A_n(T)}{v^n} \tag{4.56}$$

Then the A_j are found by a curve-fitting process, care being taken to extend the data to sufficiently low pressures for $A_1(T)$ to be equivalent to the real second virial coefficient $B(T)$.

The determination of the interatomic potential parameters at a number of different temperatures does not yield a generally valid potential. Different values of these parameters may be found by analysis of the data at other values of T. However, if the analytical form is a reasonable facsimile of the true potential under the conditions of the experiment, the variation of these parameters over a large range of temperature should not be too great.

4.15 RADIATION DAMAGE INTERATOMIC POTENTIALS

It is tempting to try to obtain an interatomic potential from experimental results concerning radiation damage in solids if we wish later to apply this potential to the theoretical interpretation of some other aspects of radiation damage. Unfortunately, as we have already pointed out, the multiple atomic collisions occurring in a solid make a simple interpretation in terms of a pair potential rather difficult. It is nonetheless feasible provided that the investigation is based on an easily measurable experimental quantity. We shall discuss here briefly four such types of data.

(1) *Focusing Energies in Crystals*

The phenomenon of focusing of kinetic energy along close-packed rows of atoms in a crystal lattice was first predicted by Silsbee [28]. It is well documented [29, 30] and will not be described in detail here. Focusing consists of a sequence of elastic collisions of atoms along a low-index direction of a crystal, where the momentum perpendicular to this direction (and hence the energy loss from the sequence) is small. This enables energy to be transported over an abnormally long distance in a crystal by comparison with an amorphous solid.

In a series of ingenious experiments, Thompson and Nelson [30, 31] investigated the preferential ejection of atoms from the surface of a single crystal in certain directions following ion bombardment of the crystal. They interpreted this ejection as being the end-product of focused collision sequences beginning at a back-scattered atom in the interior of the crystal and impinging on the surface. By measuring the energy of the atoms ejected they obtained a *focusing energy*, or maximum energy of propogation of a *focuson*, as a focused collision sequence is usually termed.

It is perhaps appropriate to add a note of caution. While it is intuitively highly probable that focused collisions do exist, this author is not at present aware of any completely *unambiguous* experimental proof of their existence. It has been suggested that the preferential ejection observed by Thompson and Nelson could be explained simply in terms of a sputtering effect in the first few surface layers, certain directions being favored by the surface geometry of the crystal. An experiment involving ejection from the back surface of an ion-bombarded crystal of thickness greater than the energetic ion range would be more conclusive from this point of view.

If we however suppose that the ejected atoms do represent the end of focusons created in the interior of the crystal, it is possible to use the focusing energy to determine the interatomic potential parameters. The collision sequence may be described in classical terms. A detailed theory has been formulated by Lehmann and Leibfried [32], who obtained an equation incorporating the focusing energy for a $\langle 110 \rangle$ focuson in an fcc crystal and the BM parameters A and B, of the form

$$2 = \frac{r_0}{2R} + 0.347 \frac{r_0}{R}\left(\frac{1}{BR}\right) - \frac{0.48 + 0.17r_0/R}{1 - 2R/r_0}\left(\frac{1}{BR}\right)^2 \qquad (4.57)$$

where

$$R = (1/2B)\ln(2A/E_f) \qquad (4.58)$$

The nearest-neighbor distance r_0 is also the distance between focusing collisions in this direction. The focusing energy E_f from experimental measurements in gold was given as 280 ± 50 eV. Since there are two BM constants some additional experimental information is required. This may take the form of the compressibility for example (see Section 4.9).

The Lehmann–Leibfried method is not unique in relating E_f to the potential. Another somewhat simpler and more approximative equation has been obtained by Thompson and Nelson [33]:

$$E_f = 2Ae^{-Br_0/2} \qquad (4.59)$$

In addition, the final values of the parameters will depend on methods of evaluating the closed shell contribution to the compressibility K_B. Thus the different theories lead to a slight variation in the potentials obtained. The BM potential suggested by Thompson has already been quoted in Eq. (4.49). A slightly different potential has been proposed by Van Vliet [34], who used the Lehmann–Leibfried result for the focusing energy E_f. This takes the form (again for gold)

$$V(r) = 3.6 \times 10^5 e^{-5.0r} \qquad (4.60)$$

where $V(r)$ is expressed in electron volts and r in angstroms.

(2) Displacement Threshold Energies

Again a property linked with the interaction between atoms, the displacement threshold energy E_d is defined as the minimum energy which must be given to a lattice atom in order to displace it permanently from its lattice site. It has been shown by both theory [17, 35] and experiment [36] to be dependent on the initial direction of motion of the atom relative to the potential barriers formed by neighboring lattice atoms. For example, the direction requiring the least energy for displacement is $\langle 100 \rangle$ for an fcc lattice and $\langle 110 \rangle$ for a simple cubic lattice. Again the problem of relating $V(r)$ to E_d may be approached from the theoretical side in an elementary or a highly complicated fashion. In very simple terms, the potential barrier formed by a ring of four atoms about the fcc $\langle 100 \rangle$ direction is given in terms of the BM potential by the equation

$$E_d^{\langle 100 \rangle} = 4Ae^{-Br_0/\sqrt{2}} \tag{4.61}$$

This formula may be used to check an already determined potential or (grouped with other experimental information) to obtain the parameters A and B. For example, substituting the parameters of Eqs. (4.49) and (4.60) into the Eq. (4.61) gives values of E_d for the $\langle 100 \rangle$ direction of gold equal to 31 eV and 56 eV respectively [34].

This simple formula is necessarily an overestimate of E_d when we consider the relaxation of the potential barrier as the focused atom passes through it. Moreover, computer simulation techniques have shown that the permanent displacement of an atom is influenced by the complicated many-body dynamics of the displacement process [35]. The use of a computer to solve the many-body equations of motion represents a more realistic method of relating the displacement threshold to the potential. It is of course necessary to assume a two-body potential before performing the simulation, but the displacement threshold energies predicted by the computer based on this assumed $V(r)$ are of use in establishing the validity of the latter. As we noted in Section 4.9, it was the predicted values of E_d which helped GGMV [17] to decide between their potentials 1, 2, and 3.

The experimental method for determining E_d consists of irradiating a precisely oriented monocrystal at low temperatures with a well-defined, monoenergetic collimated beam of electrons in the kilovolt energy range and measuring the energy threshold for onset of radiation damage by a resistivity method. The classical mechanics of energy transfer then permits an estimation of E_d. For Au, Bauer and Sosin [36] obtained the value $E_d = 35$ eV for displacement of atoms in the $\langle 100 \rangle$ direction.

Because of the dynamical aspects of the problem outlined above, it is difficult to use the displacement threshold to determine *a priori* the form or the parameters of the interatomic potential. It does, however, provide a reasonable check on a potential derived using other experimental information.

(3) Ranges of Energetic Atoms in Solids

This is not strictly speaking a radiation damage topic since we are interested in the bombarding particle rather than its effects on the stopping medium. The complications of the many-body effects are again evident. There can be a considerable spread in experimental ranges, which for the low kilovolt energy particles are of the order of a few tens of angstroms. Care must be taken in experiments on single or polycrystals to avoid bombarding in low-index crystal directions, which would lead to anomalous penetration effects such as channeling and the directionally enhanced diffusion "supertail" phenomenon [37].

Assuming an amorphous solid it is possible to derive an expression for the mean total path length of an ion slowing down in a solid from its initial energy to the displacement threshold energy, in terms of the BM parameters [38]. Alternatively, a computer simulation approach may be used to obtain a statistical mean range, again based on a suitable potential, and the results compared with experiment [39]. It was in fact the computer code developed for this purpose by Robinson and Oen [39] which aroused the recent widespread interest in the channeling phenomenon when it began to follow a penetrating ion over an anomalously long distance in the crystal.

The same comment applies to range determinations as to the displacement threshold. Their use lies chiefly in checking whether or not an already established potential is reasonable. Neither theory nor experiment is sufficiently accurate to establish satisfactory limits for the potential parameters. It is possible to use experimental ranges and a simple theoretical model to investigate the energy regime in which a certain form of potential is valid by comparing predicted path lengths and experimental values at different energies. In this way it might be possible to find the value of r at which a screened Coulomb or a BM potential deviates from a realistic form of interaction.

(4) Channeling

Instead of choosing the orientation of a crystal subjected to an ion beam bombardment so that it becomes equivalent to an amorphous solid, we might use the crystal lattice to channel the ions through a thin film and measure their energy loss on transmission. The channeled ions undergo mainly glancing collisions with the atoms of the walls of the channel. The impact parameters for the most perfectly channeled particles are determined by the channel dimensions. One might, therefore, suggest that an expression could be developed for the energy loss in terms of the interatomic potential and crystal thickness. Thus the potential parameters could be matched to experiment.

There is a complication which arises from the electronic collisions of the channeled ion. Its energy will be in the range ~ 10–100 keV or greater to ensure transmission through crystals which need not be unreasonably thin. Since the electron density is high in the center of the channels, its inelastic energy loss will be more than usually significant, tending to mask the relatively low elastic energy loss to atoms of the wall of the channel. Thus when seeking an elastic pair interaction we have to contend with a small signal/noise ratio.

It is now well established that the electronic energy loss shows quite a strong oscillatory dependence on the atomic number Z of the channeled ion, with minima at $Z = 10$, 32, and 51, independent of the Z of the stopping crystal [40]. This is a shell effect and has been successfully accounted for [41] on the basis of the Firsov semiclassical theory of inelastic energy loss, which related the latter to the impact parameter [42].

Nelson has suggested that the interatomic potential should show similar shell effects [43]. Because of the predominance of electronic losses in transmission channeling experiments, he used the back-scattering of channeled ions, measuring the reduction in scattering yield as the $\langle 110 \rangle$ axis of an Au single crystal was rotated about an incident beam of energetic ions. The ratio of dip to shoulder height in the angular dependence of back-scattered yield for varying incident ion Z exhibited a tendency towards oscillatory behavior. The first minimum this time was at $Z = 13$.

A simple theoretical investigation of the effects of overlap of different electron shells on the interatomic potential has been performed by Cheshire and Poate [44]. They used a simple Hartree form of pair potential which incorporates shell effects in the electron density term (see Eq. (2.12)). This was compared with a Bohr screened Coulomb potential evaluated for impact parameters in the back-scattering range and for different values of Z. The result of their calculation is reproduced in Fig. 4.12. It was found that the variation with Z of the ratio of the two potentials at different fixed separations followed an oscillatory curve similar to that of Nelson's experimental results.

It therefore appears that channeling may be used to obtain some information on the nature of $V(r)$, and in particular on the effects of overlap of successive electron shells of the interacting atoms. However, it should be pointed out that the interpretation of back-scattering experiments is not entirely unambiguous since, for example, dechanneling effects could mask a potential variation. It is early to say whether the significant theoretical and experimental difficulties may be sufficiently overcome to define the potential to a satisfactory degree of precision.

4.16 SHOCK WAVE EXPERIMENTS

When a strong shock wave is transmitted through a solid it carries in front of it a highly compressed region in which the atomic volume may be reduced by a factor of two. Therefore in this region there must be a considerable degree of overlap of the closed electron shells during atom–atom interactions. Thus it should be possible to interpret experimental data concerning the shock wave in such a way as to obtain information on the interatomic potential. Such a study was initiated by Duvall and Koehler [45] for the purpose of obtaining a potential for the noble metals.

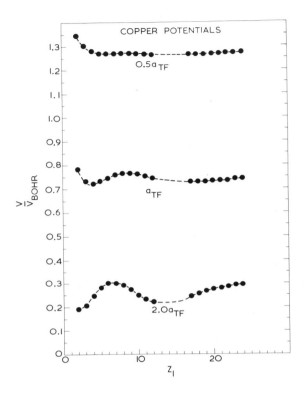

Fig. 4.12. Variation of the ratio of the Hartree potential (V) to a Bohr screened Coulomb form (V_{Bohr}) with atomic number Z_1 for different impact parameters of the order of the Firsov screening length a_{TF}. (After Cheshire and Poate [44].)

It is possible to relate the pressure and density of the solid both in front of and behind the shock wave to the measurable shock velocity and particle velocity behind the sharp shock front. The change in volume may then be used to fit an interatomic potential of the form

$$W(v) = A\left(\frac{v_0}{v}\right) + B\left(\frac{v_0}{v}\right)^{2/3} - C\left(\frac{v_0}{v}\right)^{1/3} + H \exp\left\{\alpha\left[1 - \left(\frac{v}{v_0}\right)^{1/3}\right]\right\} \quad (4.62)$$

where v and v_0 express the atomic volume under compression and at zero pressure. In this equation the first three terms are concerned with the static electron gas compressibility and the last term represents a closed shell repulsion of the BM type. The pressure associated with a given volume change is given by

$$P_0 = -\partial W/\partial v \quad (4.63)$$

This pressure P_0 is an isothermal quantity applying to the solid at $0°K$, and is therefore not directly measurable. It is necessary to reduce the experimental data for different temperatures to a zero temperature condition. This is achieved using the Gruneisen equation of state, where the Gruneisen constant is assumed to be a function of volume only. Since there are several possible expressions for this constant, the reduction of the pressure–volume data to $0°K$ must introduce a major source of uncertainty into this type of potential determination.

Assuming that the value of P_0 is known for several different pressures, the five potential constants A, B, C, H, and α are obtained from three points on the $P_0 - v$ curve, together with $W(v_0)$ (related to the crystal cohesive energy), and the fact that $P_0(v_0) = 0$. Duvall and Koehler obtained values of these constants for Cu, Ag, and Au. Using their full potential energy W to calculate the compressibility, they found surprisingly good agreement with the measured compressibility extrapolated to $0°K$ in the case of Cu, but less satisfactory for the other two metals.

At the present time it is difficult to assess the value of the shock wave approach to interatomic potentials. In its favor is the fact that it permits a study of $V(r)$ over a much wider range of r than that applying to equilibrium crystal data (pressures of more than 1000 kbar can be achieved). This is counterbalanced by the difficulty mentioned above concerning the extrapolation of the $P - v$ data to $0°K$, as well as by the increasing significance of many-body atomic interactions at these high pressures.

4.17 CONCLUSION

In this chapter we have considered a wide range of interatomic potentials based on mathematical approximations to the forces acting between atomic systems. The only common aspect of these potentials is that their variation with distance may be written as an analytical function, as distinct from an integral equation or a numerical table. They are consequently amenable to theoretical applications, at least in principle. The simpler forms may be used in analytical theories while the more complicated forms, such as the composite polynomial potentials, often require the use of a computer when applied to atomic problems.

Empirical potentials cover the complete range of interatomic separation from very high energy collisions where the nuclear repulsion plays a dominant role to distances greater than the second-neighbor separation in crystals. There is no one form of interaction applicable to all interatomic distances, as most forms lose their validity at either small or large r depending on their theoretical base and parameter values. The choice, therefore, depends on the problem under consideration.

It is admittedly a broad generalization to state that the potential is relatively well known in two regions of interatomic separation. First, when the atoms are sufficiently close together ($r \lesssim 0.1$ Å) the interaction is assumed to be that of two nuclei screened by the electron charge cloud. Second, when the distance is approximately equal to or greater than that of the nearest-neighbor separation in the solid, and the closed shells of orbital electrons may be considered to be just touching each other, crystal data may be used to determine constant parameters for different analytical forms. In addition, the theory of pseudopotentials provides a means of obtaining a two-body interaction at the equilibrium separation in the crystal when applied to certain metals. This will become evident in the following two chapters.

There remains the awkward range of separation between these two extremes, where the many-body effects of overlapping electron shells hinder a simple theoretical approximation and where experimental data are scarce. The large-r and small-r forms may be extended to cover this range, although there is no unique and reliable method of performing this extrapolation. It is unfortunately the case that the nature of the interatomic potential at these intermediate separations is of vital importance for many atomic collision problems such as that of radiation damage in solids. On the basis of present knowledge the best potential for this range would perhaps involve a compromise between some variant of the TF interaction at small r and the BM form with parameters evaluated at equilibrium separations in the solid. Some methods of joining these two forms of potential were discussed in Section 4.10.

REFERENCES

1. M. Born and K. Huang, "Dynamical Theory of Crystal Lattices." Oxford Univ. Press, London and New York, 1954.
2. J. O. Hirschfelder, C. F. Curtiss, and R. B. Bird, "Molecular Theory of Gases and Liquids," Chapter 1. Pergamon Press, Oxford, 1967.
3. R. A. Buckingham, *Proc. Roy. Soc.* **A168**, 264 (1938).
4. R. A. Buckingham, *J. Plan. Space Sci.* **3**, 205 (1961).
5. J. E. Lennard-Jones, *Proc. Roy. Soc.* **A106**, 463 (1924).
6. J. E. Lennard-Jones, *Proc. Roy. Soc.* **A106**, 441 (1924)
7. P. M. Morse, *Phys. Rev.* **34**, 57 (1929).
8. J. W. Linnett, *Trans. Faraday. Soc.* **36**, 1123 (1940).
9. L. A. Girifalco and V. G. Weizer, *Phys. Rev.* **114**, 687 (1959).
10. N. Bohr, *Kgl. Dansk. Vid. Selsk. Mat.-Fys. Medd.* **18**, No.8 (1948).
11. J. A. Brinkman, *J. Appl. Phys.* **25**, 961 (1954).
12. M. Born and J. E. Mayer, *Z. Phys.* **75**, 1 (1932) *et seq.*
13. H. B. Huntington, *Phys. Rev.* **91**, 1092 (1953).
14. K. Fuchs, *Proc. Roy. Soc.* **A153**, 622 (1936).
15. H. B. Huntington, *Solid State Phys.* **7**, 213 (1958).
16. W. C. Overton, Jr. and J. Gaffney, *Phys. Rev.* **98**, 969 (1955).

17. J. B. Gibson, A. N. Goland, M. Milgram, and G. H. Vineyard, *Phys. Rev.* **120**, 1229 (1960).
18. J. A. Brinkman, AEC Rep. NAA-SR-6642 (1960) or Rendiconti della Scuola Internazionale di Fisica 'E. Fermi,' p. 830 (1960).
19. G. Leibfried, Rendiconti della Scuola Internazionale di Fisica 'E. Fermi,' p. 227 (1960).
20. J. P. Genthon, CEA (France) Rep. No. R-3712 (1969).
21. M. T. Robinson, *Phys. Rev.* **179**, 327 (1969).
22. M. W. Thompson, *Phil. Mag.* **18**, 377 (1968).
23. R. A. Johnson, *Phys. Rev.* **134**, A1329 (1964).
24. R. A. Johnson, *Acta Mett.* **12**, 1215 (1964).
25. R. A. Johnson, *Radiat. Effects* **2**, 1 (1969).
26. A. Englert, H. Tompa, and R. Bullough. Fundamental Aspects of Dislocation Theory, p. 273, NBS Special Publ. 317 (1971).
27. R. Bullough and R. C. Perrin, "Radiation Damage in Solid Materials," p. 233. IAEA, Vienna, 1969.
28. R. H. Silsbee, *J. Appl. Phys.* **28**, 1246 (1957).
29. L. T. Chadderton, "Radiation Damage in Crystals." Methuen, London, 1965.
30. M. W. Thompson, "Defects and Radiation Damage in Metals." Cambridge Univ. Press, London and New York, 1969.
31. R. S. Nelson, "The Observation of Atomic Collisions in Crystalline Solids." North-Holland Publ., Amsterdam, 1968.
32. Chr. Lehmann and G. Leibfried, *Z. Phys.* **162**, 203 (1961).
33. M. W. Thompson and R. S. Nelson, *Proc. Roy. Soc.* **A259**, 458 (1961).
34. D. Van Vliet, Thesis, Univ. of Cambridge (unpublished); see also L. T. Chadderton and I. M. Torrens, "Fission Damage in Crystals," p. 249. Methuen, London, 1969.
35. I. M. Torrens and L. T. Chadderton, *Phys. Rev.* **159**, 671 (1967).
36. W. Bauer and A. Sosin, *Phys. Rev.* **136**, 255 (1964).
37. J. A. Davies, L. Eriksson, and J. L. Whitten, *Canad. J. Phys.* **46**, 573 (1968).
38. D. K. Holmes and G. Leibfried, *J. Appl. Phys.* **31**, 1046 (1960).
39. M. T. Robinson and O. S. Oen, *Phys. Rev.* **132**, 2385 (1963).
40. F. H. Eisen, *Canad. J. Phys.* **46**, 561 (1968).
41. I. M. Cheshire, G. Dearnaley, and J. M. Poate, *Proc. Roy. Soc.* **A311**, 47 (1969).
42. O. B. Firsov, *Zh. Eksperim. Teor. Fiz.* **36**, 1517 (1959) [English transl.: *Sov. Phys. JETP* **9**, 1076 (1959)].
43. R. S. Nelson, *Phys. Lett.* **28A**, 676 (1969).
44. I. M. Cheshire and J. M. Poate, *in* "Atomic Collision Phenomena in Solids," p. 351. North-Holland Publ., Amsterdam, 1970.
45. G. E. Duvall and J. S. Koehler, USAEC Rep. No. 4376 (1959).

CHAPTER V

PSEUDOPOTENTIAL THEORY

It would be well to apologize in advance for the gaps which will inevitably be present in a chapter of so general a title as the present. The aims are, however, a little more specific than the chapter heading suggests. We wish to achieve a good conceptual grasp of the principle of the pseudopotential method, in order to understand how it may be applied to interactions between atoms, and what are its advantages and limitations. The chief aim in this book is to show how a two-body interatomic potential may be developed from pseudopotential theory, although it is hoped that this and the following chapter will serve as an elementary review of the present state of the theory. The general principles will be outlined in the present chapter and particular types of pair interactions based on the pseudopotential method will be discussed in Chapter VI.

5.1 WHAT IS A PSEUDO-ATOM?

For many years it was known that quite good quantitative treatments of some properties of metals could be obtained from the application of the free-electron theory, in which the electrons were presumed to roam through the metal at will. What was not evident was why the ions had so little influence on these electrons. The ion core of a metal atom contains strong force fields which must greatly perturb a conduction electron penetrating it, in apparent contradiction of the free electron theory. Nevertheless the old nearly free electron method of Mott and Jones [1] had a strong measure of success in predicting metallic properties without a great deal of theoretical justification.

To a certain extent (as will become evident in what follows) pseudopotential theory vindicates the nearly free electron assumption, puts the theory on a firmer physical footing, and provides a new mathematical formalism for the calculation of many properties of metals. It should be stressed that it is by no means the ultimate in the theory of metals since it may be successfully applied only to the "normal" metals, that is those whose ion cores do not overlap in the crystalline state. Attempts are currently under way to extend its application to the transition and noble metals, but their success is as yet rather limited.

Let us consider what forces an electron experiences in the neighborhood of a metal ion. The strong Coulomb attraction of the ion core is opposed by the repulsion due to the operation of the Pauli principle for the electrons of the closed shells, as we have seen in Chapter I. In the free atom the outermost electrons would distribute themselves in such a way that these two contributions tend to cancel each other, neglecting for the present any kinetic energy and exchange phenomena. In a metal the conduction electrons tend to move in regions of minimum ionic potential, and their accumulation forms what is known as a screening charge, which must balance the ionic charge in order to satisfy high conductivity requirements. The net effective interaction experienced by an electron as a result of the cancelation of the two principal contributions is quite small, and it is this interaction which is known as a *pseudopotential*. The ion in a metal taken together with its screening charge has been called a *pseudo-atom* [2]. Such pseudo-atoms are the units of a metal in the pseudopotential treatment and their effective scattering potential is small.

We shall now review, using a minimum of mathematics, the principles of pseudopotential theory. For a much more extensive and rigorous treatment of the subject the reader is referred to the book by Harrison [3].

5.2   Orthogonalized Plane Waves

The physical basis of the theory is the exclusion of the conduction electrons of the metal from the region of space occupied by the ion cores, which then form the nuclei of the pseudo-atoms. This concept may be represented using the technique of *orthogonalized plane waves* (OPW), in which the conduction electron wave functions are forced orthogonal to any core state. If we use the normal notation where $|\mathbf{k}\rangle$ represents the free electron wave function $\Omega^{-1/2}e^{i\mathbf{k}\cdot\mathbf{r}}$ (Ω being the volume of the system), and let $|\alpha\rangle = \psi_\alpha(\mathbf{r})$ represent a core state, then an OPW may be written

$$\chi_k = |\mathbf{k}\rangle - \sum_\alpha \langle\alpha|\mathbf{k}\rangle \, |\alpha\rangle \tag{5.1}$$

Then χ_k is orthogonal to any core state $|\beta\rangle$ since

$$\langle\beta|\chi_k\rangle = \langle\beta|\mathbf{k}\rangle - \sum_\alpha \langle\alpha|\mathbf{k}\rangle\langle\beta|\alpha\rangle = 0 \tag{5.2}$$

We may now write the conduction band wavefunction ψ_k as a function involving a linear combination of orthogonalized plane waves based on the reciprocal lattice vectors $\mathbf{q}$:

$$\psi_k = \sum_\mathbf{q} a_\mathbf{q} \chi_{\mathbf{k}+\mathbf{q}}$$
$$= \sum_\mathbf{q} a_\mathbf{q}|\mathbf{k}+\mathbf{q}\rangle - \sum_{\mathbf{q}\alpha} \langle\alpha|\mathbf{k}+\mathbf{q}\rangle \, |\alpha\rangle \tag{5.3}$$

This may be simplified by introducing a projection operator P, which projects onto the core states:

$$P = \sum_\alpha |\alpha\rangle \langle\alpha| \tag{5.4}$$

Then ψ_k may be written

$$\psi_k = \sum_\mathbf{q} a_\mathbf{q}(1 - P)|\mathbf{k}+\mathbf{q}\rangle = (1 - P)\phi_k \tag{5.5}$$

If we substitute this into the Schrödinger equation for the energy eigenstates,

$$H\psi_k = E_k\psi_k \tag{5.6}$$

a similar equation in ϕ_k ensues:

$$\{H + (E_k - H)P\}\phi_k = E_k\phi_k \tag{5.7}$$

As can be seen by inspection, this is equivalent to the Schrödinger equation with the potential energy $V(r)$ replaced by a term

$$W = V(r) + (E_k - H)P \tag{5.8}$$

This term W is known as the *pseudopotential* and the function ϕ_k is the *pseudo-wavefunction*. The second term in W is repulsive and tends to cancel the strongly attractive $V(r)$, leaving only a weak attractive field whose Fourier components may be easily calculated. We note that since P and E_k enter the pseudopotential of Eq. (5.8), W is in operator form, so that in any full treatment it should be handled in a self-consistent fashion. However, in many cases it has been found a convenient and reasonable approximation to substitute a simple nonoperator term V_p for $(E_k - H)P$:

$$W(r) = V(r) + V_p \tag{5.9}$$

This is known as a *local* pseudopotential, being a function of position only.

5.3 THE PSEUDOPOTENTIAL FOR A SYSTEM OF IONS

If we consider any assembly of N ions constituting the system presented in the previous section, Eq. (5.8) now represents the general pseudopotential of the system. Then $V(r)$, the potential of the ion cores, may be written as a sum of individual potentials of each ion. If $(E_k - H)$ depends only on core energy eigenvalues and angular momentum, and if the core state $|\alpha\rangle$ depends on the position of the ion, it will be possible to separate the pseudopotential W into individual ionic pseudopotentials, spherically symmetric and overlapping:

$$W(r) = \sum_i w_i(r - r_i) \tag{5.10}$$

Note that we have not implied the local approximation for the pseudopotential of this equation.

The interaction between an electron and a system of ions depends on the plane wave matrix elements of the pseudopotential $\langle k + q | W | k \rangle$, and owing to the weakness of the potential may be found by a perturbation expansion. In the *local* approximation it becomes the Fourier transform of $W(r)$:

$$W(q) = (1/\Omega) \int_\Omega W(r) e^{-iq \cdot r} \, d\tau \tag{5.11}$$

We can then write $W(q)$ as the sum of individual ionic contributions:

$$W(q) = (1/\Omega) \sum_i \int_\Omega w(r - r_i) e^{-iq \cdot r} \, d\tau \tag{5.12}$$

$$= S(q) w(q) \tag{5.13}$$

where

$$S(q) = (1/N) \sum_i e^{-iq \cdot r_i} \tag{5.14}$$

$S(q)$ is known as the *structure factor* of the system and depends only on the ion positions. The other factor $w(q)$, given by

$$w(q) = (1/\Omega_0) \int w(r) e^{-iq \cdot r} \, d\tau \qquad (\Omega_0 = \Omega/N) \tag{5.15}$$

is known as the *form factor*, which is the Fourier transform of a single ionic pseudopotential, independent of the ion positions.

This is an extremely useful type of factorization. For instance, if we are concerned with problems involving imperfections or impurities in crystals, we need only consider modification of the structure factor to a first approximation. The perturbation of the electron wavefunctions in the region of the imperfection could affect the form factors of the relevant ions and render the problem more intractable. This would be most significant in crystals where the closed shells of the ions touch or overlap, in which case the pseudopotential formulation itself falls into question.

5.4   The Structure Factor for a Periodic Lattice

Having separated the pseudopotential according to Eq. (5.13), let us now look briefly at the part which depends on the configuration of the ions in the system. The structure factor for any system is, according to Eq. (5.14), a sum of sinusoidally varying terms depending on the possible electron wavenumbers and ion positions. For a perfect periodic lattice the Born–Von Karman boundary conditions imposed on a box containing the ion system limit the possible values of $\mathbf{q}$:

$$\mathbf{q} = \sum_{n=1}^{3} (2\pi m_n / L_n)\hat{\mathbf{a}}_n \tag{5.16}$$

where L_n are the dimensions of the box, $\hat{\mathbf{a}}_n$ the unit reciprocal lattice vectors, and m_n is an integer between 1 and the number of ions in the nth direction. It is a simple exercise to show that for such a lattice $S(q)$ is zero everywhere except at lattice wavenumbers, where it is unity. It is this property which leads directly to the Bragg condition for electron diffraction, the electron always changing its wavenumber by a discrete lattice wavenumber.

The presence of a defect in the lattice—for example a vacancy—will alter the structure factor. If we neglect lattice relaxation about the vacancy, the $S(q)$ remains equal to unity at the lattice wavenumbers, but where it was previously zero it becomes $1/N$. This gives rise to the physical phenomenon of scattering by the defect.

In similar fashion the structure factor for a lattice whose ions are slightly displaced from their lattice sites, for example through lattice vibrations, may be calculated by substituting $\mathbf{r}_i + \delta\mathbf{r}_i$ for $\mathbf{r}_i$ and expanding about the $S(q)$ for the perfect lattice:

$$S(q) = \frac{1}{N}\sum_i e^{-i\mathbf{q}\cdot\mathbf{r}_i}\left\{1 - i\mathbf{q}\cdot\delta\mathbf{r}_i - \frac{(\mathbf{q}\cdot\delta\mathbf{r}_i)^2}{2} + i\frac{(\mathbf{q}\cdot\delta\mathbf{r}_i)^3}{6} + \cdots\right\} \tag{5.17}$$

In this case the first-order terms in conjunction with the form factor lead to the electron–phonon interaction, and the first- and second-order (harmonic) terms may be used to calculate the vibration spectrum. The third-order (anharmonic) terms represent the phonon–phonon interaction. The first two terms of such an expansion may be used to calculate the change in structure factor brought about by the neighboring atoms around a defect in the crystal. Thus from (5.17)

$$\delta S(q) = (-i\mathbf{q}/N) \sum_i \delta \mathbf{r}_i \, e^{-i\mathbf{q}\cdot\mathbf{r}_i} \qquad (5.18)$$

Knowing or estimating the relaxed positions of the neighboring atoms, one might use in this manner the modified structure factor to calculate how much of the resistivity due to the vacancy arises from local distortion of the lattice.

Before leaving the subject of structure factors it should be noted that the value of $S(q)$ for a lattice containing more than one ion per unit cell can be less than unity at the integral wavenumbers. This is simply a question of geometry, and does not alter its significance in any way. For simple lattices then, the unit value of the structure factor means that the Fourier transform of the total pseudopotential is simply the form factor of the ions, which we shall now go on to consider.

5.5   THE SINGLE ION FORM FACTOR

The form factor which we derived from Fourier transformation of the individual ion pseudopotential in Eq. (5.15) is a simplified one, based on a local pseudopotential. In the nonlocal operator form it would be energy-dependent and would be given by the matrix element $\langle \mathbf{k} + \mathbf{q} | w(\mathbf{r} - \mathbf{r}_i) | \mathbf{k} \rangle$. When $w(r)$ is not an operator the $\mathbf{k}$ cancels out and we are left with $w(q)$, as in Eq. (5.15), depending only on $\mathbf{q}$. The full form factor can be calculated from first principles for matrix elements between states on the Fermi surface (where $|\mathbf{k}| = |\mathbf{k} + \mathbf{q}| = k_F$), and is known as the OPW form factor. Several of these, together with the method of calculation, are given by Harrison in his book [3] and are illustrated in Fig. 5.1. The small q section of the curves is primarily influenced by the screened Coulomb attraction and at large values of q the repulsive part of the potential becomes important.

Since the calculation of the nonlocal form factors is rather involved a number of semilocal and local approximations have been made. Perhaps the most often quoted of these is that of Heine and Abarenkov [4]. They constructed a sphere about the ion of radius R_m, outside which the potential of the ion was purely Coulombic. Bearing in mind that what was required for a model of the ion was in fact the potential as seen by a conduction electron

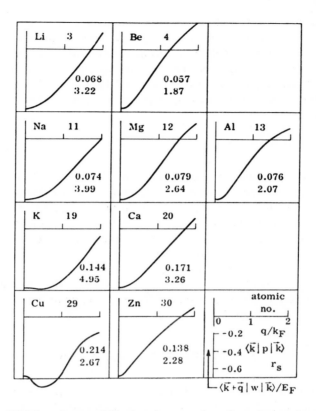

Fig. 5.1. OPW form factors of Harrison for a number of metals, in units of the Fermi energy E_F, plotted against q/k_F. The square at the bottom right of the figure contains the key. $\langle \mathbf{k} | p | \mathbf{k} | \rangle$ and r_s are connected with the calculation of the nonlocal pseudopotential) (see Harrison [3]). Letters with arrows correspond to boldface in text. (After Harrison [3].)

outside the ion core rather than the true potential function, which was rapidly varying, they represented the bare ion potential by the form (see Fig. 5.4(a))

$$w_b(r < R_m) = \sum_l A_l(E) P_l$$

$$w_b(r > R_m) = -(Ze^2/r)$$

(5.19)

Here A_l is a parameter which varies slowly with E, the energy of the incident conduction electron, and with R_m; P_l is a projection operator which extracts from the incident wave function the component with angular momentum l; $A_l(E)$ is fitted to the spectroscopically obtained energy levels of the free ion. Therefore, the model ion potential has eigenfunctions which are the same as those of the true potential outside the core, although they are much too slowly varying inside the ion. The strong ionic potential has been replaced by a much weaker one (and consequently much easier to handle) which behaves in a similar manner outside the ion core, so that in effect the conduction electrons do not notice the difference since they do not penetrate the core. In practice, Heine and Abarenkov used only three "constants" $A_0(E)$, $A_1(E)$, and $A_2(E)$, and made $A_l(E)$ equal to $A_2(E)$ for larger values of l. Then since $\sum P_l = 1$,

$$w_b(r < R_m) = -A_2 - (A_0 - A_2)P_0 - (A_1 - A_2)P_1 \tag{5.20}$$

Animalu and Heine [5] have computed the model potential for a number of elements. Using $A_l(E_F)$, their form factors for comparison with those of Harrison are illustrated in Fig. 5.2. We shall be returning to the Heine–Abarenkov–Animalu (HAA) model potential in greater detail in Chapter VI.

The HAA model potential has been modified in some recent work by Shaw [6, 7] in several ways. First, terms were included only for angular

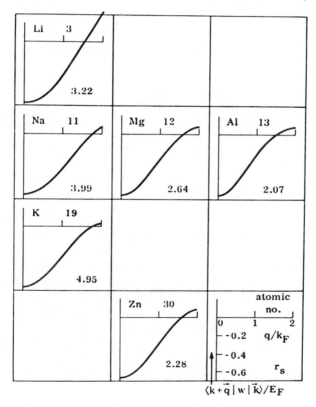

Fig. 5.2. OPW form factors of HAA for several metals. The scale is explained in Fig. 5.1. Letters with arrows correspond to boldface in text. (After Harrison [3].)

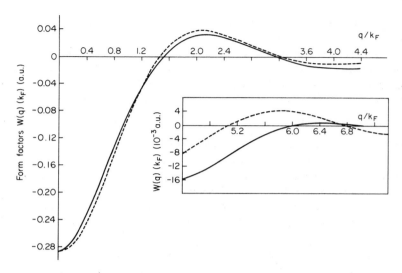

Fig. 5.3. Optimized model OPW form factor (———) for Al compared with the form factor of HAA (– – –). (After Shaw [7].)

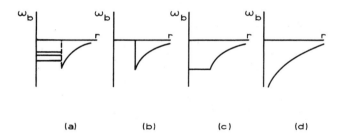

Fig. 5.4. Schematic comparison of bare ion potentials $w_b(r)$ for (a) HAA, (b) Ashcroft, (c) Ho, and (d) Point ion models.

momentum values found in the core states, and the ion core radius was no longer considered fixed, being a function of both l and E. With this prescription, the model potential could be optimized by a variational procedure. Further corrections were included in the Shaw optimized form to allow for the electron charge density depletion inside the ion cores. The Shaw form factor for Al is compared with the normal HAA form factor in Figure 5.3.

A further simplification of the bare ion potential, which abandoned all energy and angular momentum dependence, was proposed by Ashcroft [8] (Fig. 5.4(b)):

$$w_b(r < R_m) = 0$$
$$w_b(r > R_m) = -(Ze^2/r)$$

(5.21)

This was essentially a purely local potential, where R_m was expected to be slightly modified from that of Heine and Abarenkov, but still in the neighborhood of the ion core radius. The calculation of the form factor for the Ashcroft bare ion potential is analytically simple. It will serve to illustrate the transformation to wavenumber space if we consider first the simple case where the radius R_m is assumed to be zero, that is the inversion of the simple Coulomb potential. We change to polar coordinates by letting $d\tau = 2\pi r^2 \, dr \cdot \sin\theta \, d\theta$ and $\mathbf{q} \cdot \mathbf{r} = qr \cdot \cos\theta$, so that from Eq. (5.15) the form factor becomes

$$w_b(q)_{\text{Coulomb}} = -(1/\Omega_0) \int_0^\pi \int_0^\infty (Ze^2/r)e^{iqr\cos\theta} \cdot 2\pi r^2 \, dr \sin\theta \, d\theta \quad (5.22)$$

$$= -(4\pi Ze^2/\Omega_0 q) \int_0^\infty \sin qr \, dr \quad (5.23)$$

We overcome the convergence problem by adding a factor e^{-pr} and letting p go to zero:

$$w_b(q)_{\text{Coulomb}} = -(4\pi Ze^2/\Omega_0 q^2) \quad (5.24)$$

A calculation of this type for nonzero R_m leads to the form factor

$$w_b(q)_{\text{Ashcroft}} = -(4\pi Ze^2/\Omega_0 q^2)\cos(qR_m) \quad (5.25)$$

The oscillatory term arises from the discontinuity in $w_b(r)$ at $r = R_m$. The physical significance of the Ashcroft form is that the cancelation of the Coulomb and repulsive potentials in the region of the core is complete. A small adjustment of R_m is equivalent (from the point of view of the scattering of conduction electrons) to a nonzero core potential or an incomplete cancelation.

Another similar form has been proposed by Ho [9]. The main difference between the Ho and Ashcroft forms is that instead of being very small (~ 0) inside R_m, the former is negative and constant (Fig. 5.4(c)):

$$\begin{aligned} w_b(r < R_m) &= -V_0 \\ w_b(r > R_m) &= -(Ze^2/r) \end{aligned} \quad (5.26)$$

Again the form factor may be obtained analytically, the calculation being this time a little more complicated, but similar in principle:

$$w_b(q)_{\text{Ho}} = -(4\pi/\Omega_0 q^2)\{V_0 \sin(qR_m)/q + (Ze^2 - V_0 R_m)\cos(qR_m)\} \quad (5.27)$$

Perhaps the simplest type of ion–electron interaction is the point ion potential, where R_m is effectively zero as in the Coulomb case, and mathematical representation of the ion is by means of a delta function (Fig. 5.4(d)):

$$w_b(r) = -(Ze^2/r) + V_0 \delta(r) \quad (5.28)$$

The analytical form factor for this potential is very simple:

$$w_b(q)_{\text{point ion}} = -(4\pi Ze^2/\Omega_0\, q^2) + V_0/\Omega_0 \qquad (5.29)$$

Its advantage lies in its mathematical convenience. It may be fitted to form factors calculated by more complicated methods such as the HAA form factors, to produce a characteristic value of the strength of the delta function V_0 for each metal. The potential thus calculated may be applied in its simplified form to atomic problems.

To summarize, the pseudopotential form factor for a bare ion contains all of the information necessary to describe that ion from the point of view of a conduction electron scattering from the ion core. An exact or even approximate expression for the core wavefunctions is not essential since the electron does not have sufficient energy to penetrate the closed shells. In the case of the simple metals where the core energy bands are well separated from the conduction band it appears to be quite adequate to simulate the ion by a small or zero constant local potential. In some other cases where conduction-core separation is not so evident a more complicated treatment such as that proposed by Heine and Abarenkov would be better. We shall again consider the subject of form factors in Chapter VI, in connection with the interatomic pair potential.

Outside the ion core ($r > R_m$) the Coulomb attraction prevails. The next problem is to see how this attraction, and consequently the ionic pseudopotential, is modified by the presence of a sea of conduction electrons.

5.6   ELECTRON SCREENING OF THE PSEUDOPOTENTIAL

In Chapter II we introduced the concept of a self-consistent field due to the interaction of the atomic electrons with the nucleus and with each other. The charge density of the electrons screens the Coulomb field of the nucleus so that each electron experiences a modified field, which is calculated in the Hartree approximation (neglecting exchange and other correlation) by an iterative process to achieve self-consistency. In a similar manner we can calculate the pseudopotential including the screening of the bare ion by the conduction electron gas. The problem of screening by an ideal electron gas has been treated by Bardeen [10] in a paper on the electron–phonon interaction, and more recently by Lindhard [11], who introduced the concept of a dielectric function to represent this screening. In this formulation the general screened ion form factor would bear a linear functional relation to the unscreened bare ion form factor:

$$\langle \mathbf{k} + \mathbf{q} \,|\, w_s(r) \,|\, \mathbf{k} \rangle = \langle \mathbf{k} + \mathbf{q} \,|\, w_b(r) \,|\, \mathbf{k} \rangle / \varepsilon(q) \qquad (5.30)$$

or in the simpler case of the local pseudopotential,

$$w_s(q) = w_b(q)/\varepsilon(q) \tag{5.31}$$

The function $\varepsilon(q)$ is the *static dielectric function* of the electron gas. In a metal where the conduction electrons form a cloud of energetic particles, the dielectric formulation is a good approximation.

Let us consider the effect of electrostatic screening on the potential of a point charge Ze, which will be equivalent to the large distance (small q) screening of the bare ion. Then

$$w_b(r) = -(Ze^2/r) \tag{5.32}$$

By Fourier transformation (see Section 5.5),

$$w_b(q) = -(4\pi Ze^2/\Omega_0 q^2) \tag{5.33}$$

A simple expression for the dielectric function near $q = 0$, where it differs appreciably from unity, is [12]

$$\varepsilon(q) \simeq 1 + 4\pi Ze^2/\Omega_0 q^2 \tag{5.34}$$

$$= 1 + q_s^2/q^2 \tag{5.35}$$

where q_s is a screening parameter:

$$q_s = \{4\pi Ze^2/\Omega_0\}^{1/2} \tag{5.36}$$

Thus the screened form factor becomes

$$w_s(q) = -(4\pi Ze^2/\Omega_0 q^2 \varepsilon(q)) = -[1/(1 + q^2/q_s^2)] \tag{5.37}$$

We can now transform this once again to find the screened ion pseudopotential in real space:

$$w_s(r) = -(Ze^2/r)e^{-q_s r} \tag{5.38}$$

which is the familiar screened Coulomb form.

To improve the model at lower values of r, a better estimate of $\varepsilon(q)$ which shows an anomalous effect arising from the screening by a degenerate non-interacting Fermi gas is given by the dielectric function (see Appendix 2):

$$\varepsilon(q) = 1 + \frac{2k_F m e^2}{\pi h^2 q^2} \left\{ 1 + \frac{4k_F^2 - q^2}{4k_F q} \ln \left| \frac{2k_F + q}{2k_F - q} \right| \right\} \tag{5.39}$$

This expression for $\varepsilon(q)$ is due to Lindhard [11], although it is known as the Hartree dielectric function owing to the nature of the model assumed for the electron gas which neglects any interaction between the electrons. Although it is continuous at $q = 2k_F$ it has a logarithmic singularity in its derivative, which becomes negatively infinite at this point. A plot of the function for aluminium shown in Fig. 5.5 illustrates the apparent insignificance of the

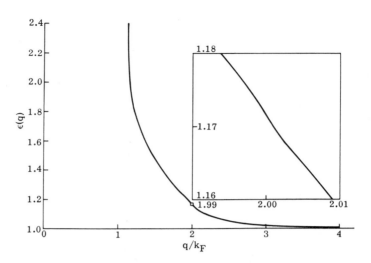

Fig. 5.5. Hartree dielectric function for Al as a function of q/k_F. The region near $q = 2k_F$ is magnified $\times 50$ in the inset. (After Harrison [3].)

singularity, which shows up as a very minor "kink" in the curve. However, this anomaly has a major effect on the Fourier transform of the form factor including the screening function, producing the so-called *Friedel oscillations* in the real space potential at large values of r. These oscillations correspond to variations in the electron charge density according to the equation [13]

$$\delta\rho \sim \cos(2k_F r)/r^3. \tag{5.40}$$

Physically they arise from the sharp cutoff of the electron energies at the Fermi surface. We shall be looking at the oscillations in greater detail when we come to consider the interatomic potential in a metal as calculated using pseudopotential theory since they play a role in the screening of the ion–ion interaction by the intervening electrons.

It is important to realize that our assumption of individual ion screening by the conduction electrons entirely depends on the linearity of the screening function. If the calculation of this phenomenon were carried out to more than first order, the screening function would no longer depend simply on the magnitude of q, and it would not be possible to give each bare ion its individual charge cloud as we have done above. It is extremely convenient to be able to represent the total pseudopotential of the ion distribution as the sum of screened ion pseudopotentials, so linear screening is a useful approximation. Of course we do not mean that physically the pseudo-atom possesses a charge cloud of its own, which would contradict the existence of the conduction electrons. The screening charge is rather made up of the averaging of effects due to a large number of conduction electrons, all of which are more or less free to roam about the metal.

The Hartree dielectric function for the noninteracting electron gas neglects exchange and correlation, and is consequently inaccurate at metallic densities. There exists a selection of corrections to the Hartree form to take the electron interaction into account, most often based on the treatment of the problem by Hubbard [14]. If we write

$$g_H(q) = \frac{2k_F m e^2}{\pi \hbar^2} \left\{ 1 + \frac{4k_F^2 - q^2}{4k_F q} \ln \left| \frac{2k_F + q}{2k_F - q} \right| \right\} \tag{5.41}$$

Hubbard showed that exchange could be included (within the perturbation formulation) by replacing $g_H(q)$ by the function

$$g(q) = \{1 - f(q)\} g_H(q) \tag{5.42}$$

where

$$f(q) = \frac{q^2}{2(q^2 + \eta k_F^2)} \tag{5.43}$$

For large q this has the effect of halving the value of $[\varepsilon(q) - 1]$, which corresponds in reality to cancelation of half of the interaction for short wavelengths.

Various methods exist for assigning a value to the parameter η. In Hubbard's original work it was simply unity. In a calculation of the form factor for the alkali metals (see also Section 6.6) Ho [9] adjusted the value of η, along with two parameters contained in the bare ion potential, to reproduce the experimental elastic constants. Another method of finding η was discussed by Geldart and Vosko [15], who related it to the ratio of the compressibilities of an interacting and a noninteracting electron gas in the static limit.

A slightly different form for $g(q)$, used by Heine and Abarenkov [4] after Sham [16] was

$$f(q) = q^2/2(q^2 + k_F^2 + q_s^2) \tag{5.44}$$

where

$$q_s^2 = 2k_F / \pi a_0 \tag{5.45}$$

and a_0 is the Bohr radius. Here q_s again plays the role of a screening parameter. A further modification included in the HAA work was a small *orthogonality correction* to allow for the charge density of a conduction state being reduced inside the ion core and correspondingly increased to preserve normalization in the rest of the volume. This had the effect of replacing the electronic charge in Eq. (5.39) and consequently in (5.42) by $e^2(1 + \alpha)$. The constant α was proportional to the volume of the core, and since the whole orthogonality correction was small, its precise value was unimportant.

An exponential form for $f(q)$ was suggested by Shaw and Pynn [16a] to extrapolate smoothly between the short and long wavelength limits of the exchange and correlation contributions:

$$f(q) = \tfrac{1}{2}\{1 - \exp(-q^2/2k_F^2) + (0.0123q^2/k_F^3)\exp(-4.374q^2/k_F)\} \quad (5.46)$$

This form is somewhat phenomenonological and is a rather complicated expression when compared with the Hubbard–Sham form.

In some recent work, Singwi and co-workers treated the short-range correlations which occur in an electron liquid at metallic densities using a local field correction expressed in terms of the pair correlation function. Their theory results in a slightly different formulation to that of Eqs. (5.42) and (5.43), namely

$$g(q) = g_H(q)/(1 - f(q)\,g_H(q)) \qquad (5.47)$$

The set of equations coupling $f(q)$, which contains the structure factor of the liquid, and $\varepsilon(q)$ may be solved numerically in a self-consistent fashion to give a dielectric function which is somewhat more realistic at long wavelengths than previous models. For details of the method and results readers are referred to the original articles [17].

In conclusion, the dielectric formulation of screening by the electron gas appears to be a reasonable approximation for a metal *provided that allowance is made for electron–electron interactions*. This may be achieved through a modification of the Hartree dielectric function along the lines of the Hubbard approximation or by a self-consistent treatment of the electron liquid. The best theoretical method presently available is probably that of Singwi *et al.*, and their numerical results or an analytical fit to these values should give the most accurate $\varepsilon(q)$ especially for large values of q. However, the Hubbard–Sham form is simpler to handle and not unrealistic in view of the other approximations involved in the treatment of many theoretical problems for which it may be used.

We have now available an expression for the screened ion pseudopotential form factor (see Eq. (5.31)) incorporating the bare ion form factor and the dielectric constant. Before going on to ask what form an interaction between two such ions will take we must first consider the total energy of an assembly of ions surrounded by their screening charge clouds.

5.7 Total Energy of an Ion–Electron System

To a first approximation this may be split up into a direct contribution due to ion–ion interactions and a contribution arising from the presence of the electrons. The former is given very simply by

$$E_{ion} = (1/2N)\sum_{i\neq j} Z^2 e^2/r_{ij} \qquad (5.48)$$

where there are N ions in the system and E_{ion} is the energy per ion. We are assuming the ions to be identical.

The repartition of the ion energy in the system may be found by a single perturbation calculation [3]. The energy of the eigenstate of wavenumber k is given by the equation

$$E(k) = \frac{\hbar k^2}{2m} + \langle \mathbf{k}|W(r)|\mathbf{k}\rangle + \sum_{q \neq 0} \frac{\langle \mathbf{k}+\mathbf{q}|W(r)|\mathbf{k}\rangle\langle \mathbf{k}|W_b(r)|\mathbf{k}+\mathbf{q}\rangle}{(\hbar^2/2m)(k^2 - |\mathbf{k}+\mathbf{q}|^2)} \tag{5.49}$$

The first term is just the kinetic energy of the electrons if there were no ions present, while the second allows the ions to exist but assumes that the electrons are completely free. We shall tend to ignore these two terms since they do not influence the ion–electron–ion interaction under constant volume conditions.

To find the total electron energy of the system it is necessary to sum the $E(k)$ of Eq. (5.49) over all possible values of k up to k_F. This sum may be converted to an integral using the standard expression for the density of states in wave number space:

$$\sum_k = (\Omega_0 N/4\pi^3) \int d^3k \tag{5.50}$$

Integrating the third term of Eq. (5.49) and factorizing it into structure and form factors, this term becomes the *band structure energy* per ion:

$$E_{bs} = \sum_{q \neq 0} S^*(q)S(q) \frac{\Omega_0}{4\pi^3} \int \frac{\langle \mathbf{k}|w_b(r)|\mathbf{k}+\mathbf{q}\rangle\langle \mathbf{k}+\mathbf{q}|w(r)|\mathbf{k}\rangle}{(\hbar^2/2m)(k^2 - |\mathbf{k}+\mathbf{q}|^2)} \tag{5.51}$$

which is usually written

$$E_{bs} = \sum_{q \neq 0} S^*(q)S(q)G(q) \tag{5.52}$$

The function $G(q)$ is termed the *energy-wavenumber characteristic*. Finally, the total electron energy per ion may be written

$$E = E_0 + E_{bs} \tag{5.53}$$

where E_0 comes from integration of the first two terms of Eq. (5.49) and is independent of the ion positions, depending only on the volume of the system.

We shall see in Section 5.9 how the band structure energy leads to an indirect ion–ion interaction via the conduction electrons, but let us first examine the energy-wavenumber characteristic in greater detail.

5.8   THE ENERGY-WAVENUMBER CHARACTERISTIC

In the case of the nonlocal operator form of the pseudopotential $w(r)$, the integral contained in $G(q)$ (see Eq. (5.51)) presents considerable difficulty. If we replace $w(r)$ by a simple analytical potential, then the matrix elements become independent of k and we are left with a form factor $w(q)$ as in Eq. (5.15). This may be taken outside the integral, which simplifies $G(q)$:

$$G(q) = (\Omega_0/4\pi^3)w(q)w_b(q) \int [d^3k/(\hbar^2/2m)(k^2 - |\mathbf{k} + \mathbf{q}|^2)] \qquad (5.54)$$

It is in fact an integral of this type which leads to the expression contained in the Hartree dielectric function $\varepsilon(q)$ (see Appendix 2). The integral may be solved by resolving the vector $\mathbf{k}$ into components parallel and perpendicular to $\mathbf{q}$ to give

$$G(q) = -\frac{\Omega_0 m k_F}{4\pi^2 \hbar^2} w(q)w_b(q)\left\{1 + \frac{4k_F^2 - q^2}{4k_F q} \ln\left|\frac{2k_F + q}{2k_F - q}\right|\right\} \qquad (5.55)$$

$$G(q) = -(\Omega_0 q^2/8\pi e^2)w(q)w_b(q)\{\varepsilon(q) - 1\} \qquad (5.56)$$

At this juncture we have not yet introduced the dielectric screening function *per se*, in relation to the screening of the bare ion. The $\varepsilon(q)$ contained in Eq (5.56) simply results from the integration of Eq. (5.54). Substituting for $w(q)$ in terms of the bare ion form factor, we now have

$$G(q) = -(\Omega_0 q^2/8\pi e^2)|w_b(q)|^2(\varepsilon(q) - 1)/\varepsilon(q) \qquad (5.57)$$

This simplified expression for $G(q)$ enables us to apply any of the simple forms of the bare ion form factor found in Section 5.5 and any variation of the linear dielectric function such as one of those mentioned in the previous section. The inclusion of an exchange contribution in $\varepsilon(q)$ does not alter Eq. (5.56), since Poisson's equation, which is used in expressing the integral in terms of $\varepsilon(q)$, is correspondingly modified.

5.9   THE PAIR INTERACTION POTENTIAL

The interaction between two ions of the ion–electron system may now be divided up into the direct Coulombic interaction, unaffected by the intervening electrons, and the ion–electron–ion interaction which we shall obtain directly from the band structure energy. Using the structure factor notation of Eq. (5.14), we express the E_{bs} given by Eq. (5.52) in the form

$$E_{bs} = (1/N^2)\sum_{q \neq 0} \sum_{\mathbf{r}} G(q)e^{-i\mathbf{q}\cdot\mathbf{r}} \qquad (5.58)$$

or

$$E_{bs} = (1/N^2) \sum_{q \neq 0} G(q) + (1/N) \sum_{r \neq 0} V'(r) \tag{5.59}$$

where $V'(r)$ represents the ion–electron–ion term in the interionic potential.

$$V'(r) = (1/N) \sum_{\mathbf{q},\, r \neq 0} G(q) e^{-i\mathbf{q} \cdot \mathbf{r}} \tag{5.60}$$

Again we convert to an integral using Eq. (5.50)

$$V'(r) = (\Omega_0/4\pi^3) \int G(q) e^{-i\mathbf{q} \cdot \mathbf{r}} \, d^3q = (\Omega_0/\pi^2 r) \int_0^\infty G(q) q \sin qr \, dq \tag{5.61}$$

Finally, adding the direct ion–ion term, the pair interaction potential becomes

$$V(r) = (Z^2 e^2/r) + (\Omega_0/\pi^2 r) \int_0^\infty G(q) q \sin qr \, dq \tag{5.62}$$

The range of the two-body interaction seems at first sight to be rather long, governed by the Coulomb term. However, upon integration of the second term it is found that the $Z^2 e^2/r$ is effectively canceled, leaving an interaction which decays much more rapidly. The general (schematic) form is illustrated in Fig. 5.6. It decreases from a repulsion at small r to an attractive minimum in the region of the nearest-neighbor distance in the crystal. At large r, oscillations occur as a result of the logarithmic anomaly in $\varepsilon(q)$. These oscillations decrease according to $\cos(2k_F r)/r^3$. An asymptotic form at large r may be calculated analytically from Eq. (5.61):

$$V(r)_{\text{asym}} = \frac{9\pi^2 Z |w_b(2k_F)|^2}{E_F} \frac{\cos(2k_F r)}{(2k_F r)^3} \tag{5.63}$$

where E_F is the Fermi energy.

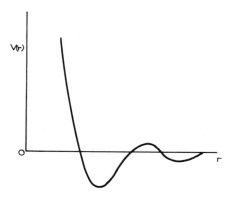

Fig. 5.6. Schematic form of the pseudopotential pair interaction.

We shall be considering particular cases of pair interactions based on pseudopotential theory in the next chapter, and shall consequently defer further discussion of the two-body potential until that point.

5.10 APPLICATION OF PSEUDOPOTENTIAL THEORY TO NOBLE AND TRANSITION METALS

The application of pseudopotential theory to a solid is limited by several factors, which have already been mentioned at various stages in the foregoing development. The small core approximation requires that there be no overlap of the closed shells of neighboring atoms, and that the core wave functions be practically identical in the free atom and in the solid. The OPW formulation is based on the nonpenetration of the ion core by the conduction electrons. Finally, the pseudopotential obtained by orthogonalizing the conduction electron wave functions to the core states must be sufficiently weak to permit the use of perturbation theory in energy calculations. All these conditions are to a greater or lesser extent satisfied by the alkali and polyvalent metals, and to date most effort in the pseudopotential field has been concentrated on these, to the virtual exclusion of noble and transition metals. In view of the evident usefulness of the theory, it is worthwhile to ask whether it might not be possible to extend it, in its present or a modified form, to these wider classes of metals.

The noble metals Cu, Ag, and Au differ from the alkali metals in that they have just completed d-shells which are large enough to produce significant overlap in the metallic state. There is therefore a modification of the wave functions from those of the free atom. The d-states should not be regarded as part of the core, but should be taken along with the conduction electrons and their wave functions computed in the pseudopotential approximation. However, these d-states are fairly localized, which means that their plane-wave expansion converges much too slowly for application of a perturbation treatment. On the other hand, if we regard the d-states as part of the core, the OPW formulation becomes suspect, since the valence electrons spend most of their time just inside the large ion cores. For the same reason there are grave doubts as to the validity of the linear screening approximation.

Notwithstanding these substantial difficulties, some recent effort has been expended towards the use of pseudopotentials for the noble metals. Nikulin and co-workers [18, 19] added a repulsion energy arising from d-shell overlap to the pseudopotential expression for the total energy. For their bare ion potential they originally chose a simple local form similar to that of Eq. (5.27), independent of energy and angular momentum. The constants were chosen from spectroscopic data, and the potential was screened in the Hubbard approximation. The d-shell overlap energy was calculated using first-order perturbation methods involving the HF electron densities in the free atom. Their values of total ground-state energy, cohesion energy, lattice constant, and compressibility of copper calculated in this way were in good agreement with experiment ($\lesssim 5\%$). The extension of their work to silver and gold yielded unsatisfactory results. Consequently they modified the bare ion potential to make it nonlocal by taking energy and angular momentum into account (up to angular momentum quantum number $l = 1$). The remaining terms were obtained as before or with slight modification. The new results this time agreed reasonably well with experiment. It was stated by Nikulin *et al.* that the theory did not hold for Au since it would have been necessary to take into account the $l = 2$ terms in the model bare ion potential.

Taking into account the very valid objections to the use of pseudopotentials for these metals and the many approximations inherent in their model, the good agreement with experiment for Cu and Ag is somewhat surprising. Nonetheless, it does provide a new method of looking at these metals and we may expect further developments along these lines which will hopefully eliminate some of the theoretical grounds for criticism.

The transition metals, because of their large incomplete d-shells, are even more difficult to handle. The d-electrons are again neither sufficiently bound to the atom for them to be unaltered in the metallic state, nor sufficiently free for them to have a slowly varying pseudowave function permitting the use of perturbation theory. The additional fact that for most of these metals the d-shells are only partly filled renders the problem quite intractable.

In a recent analysis, Harrison [20] has attempted to develop pseudopotential form factors for the transition metals. His method is to generalize the pseudopotential by reformulating it in a way which would include the atomic nature of the d-states. His true wave function in the metal is a sum of plane-wave conduction electron functions, the free atom core states and the free atom d-states. Since this last term would not give the eigenstates of the metal Hamiltonian, he includes a potential difference Δ between the free atom and the atom in the metal, which plays the role of an s–p hybridization term. The d-states can then be grouped with the core states and the pseudopotential formulated in the normal way, with the inclusion of the potential difference parameter. There is no allowance for overlap of neighboring atom d-states in this model except insofar as the energy difference parameter takes this into account.

Following the construction of an equation for the transition metal pseudopotential, a perturbation expansion is applied as in the case of the simple metals, giving the energy to second order in the pseudopotential. The screening of the potential in order to find the total energy requires a summation over all occupied states, and here an additional complication is introduced by the scattering resonances. The analysis, though involved, is feasible, and finally an expression for the total energy is developed which includes free electron and band structure contributions as before. Harrison illustrates the application of the method by calculating the pseudopotential form factor in a semilocal approximation for Cu considered as a transition metal with a filled d-band. It should be noted that his screening calculation involves a summation over states which is carried out only for completely filled or completely empty d-bands. Thus the method is strictly speaking applicable only to noble or alkaline earth metals, although it may be a good approximation for almost full or almost empty d-shells. In theory the analysis could be carried through to calculate an energy-wavenumber characteristic for these metals and subsequently a pair interaction potential, although this has not yet been completed.

This development by Harrison of the transition metal pseudopotential is an elegant treatment of the problem of a large d-band which is really neither conduction nor core. However, we are aware that a completely theoretically derived pseudopotential for the *simple* metals has some difficulty in reproducing experimental results. It is unfortunately to be expected that the disaccord will be exaggerated in this much more complicated analysis. The theoretical approach is nevertheless extremely useful. The development of the simple metal pseudopotential from first principles inspired the model and phenomenological potentials containing experimentally adjustable parameters. In the same way, it is not unlikely that a model potential for the transition or noble metals could include some features of the Harrison approach.

5.11   SUMMARY OF THE PSEUDOPOTENTIAL METHOD

In this chapter we have described the development of a method of treating the simple metals which greatly facilitates the treatment of electronic or atomic properties. The aspect of pseudopotential theory which interests us most in this book is obviously the possibility of deriving a form of real-space two-body interatomic potential. But we must realize that this is only a small part of the theory, and that many of the problems which we could tackle using this pair potential could in principle be solved in a wavenumber space approach using the appropriate form factors and structure factors. In practice though, this is often rather difficult, especially if structure factors are modified through atomic displacements. In the next chapter we shall discuss some of the pair interactions which have been derived on a pseudopotential basis, so it would be appropriate at this stage to summarize the basic assumptions and procedures which go to make up the theory.

The starting point of the theory is the small core approximation; i.e., the ion cores do not touch or overlap in the solid, and their wavefunctions are unaltered in going from the free atom to the metal. It is this assumption which limits the metals to which we may apply the basic theory. The small core hypothesis leads directly to the exclusion of the conduction electrons from the region of the core, which enables us to use the method of orthogonalized plane waves to describe the conduction electron–ion system. The conduction electrons are assumed to be described by plane waves as in the free electron theory, and these waves are orthogonalized to the core functions. When the OPW's are substituted into the Schrödinger equation of the system an equivalent equation results, with the strong true potential energy term replaced by a much weaker pseudopotential, and an accompanying pseudowave function which may be handled by perturbation theory. The pseudopotential is in the exact theory an energy-dependent operator, but it is often more convenient and does not introduce too much error to assume a local non-operator form.

The next step—a very important one for application of the theory—is the division of the pseudopotential of the system into individual ion pseudo-potentials, and factorization of its matrix elements into a form factor in-dependent of the ion positions and a structure factor depending only on the ion positions. The latter is normally simple and easily determined for periodic lattices, even if a defect is present.

The form factor is the Fourier transform of the ion core potential in the local approximation, although it is energy-dependent in the nonlocal theory. It may be calculated from first principles in the full theory, or else a form may be assumed for the bare ion potential and this transformed. Such potentials are known as model potentials, and are usually of some simple form. For example there might be an ionic radius assumed, outside which the potential is Coulombic and inside which it might be zero, a constant negative value, or a potential containing constants fitted to spectroscopic data. The exact form is not very important since the conduction electrons are excluded from the ion core and are therefore only sensitive to the external Coulombic attraction. It should however contain some parameters such as the ionic radius which are adjustable to the experimental data, since these influence the final screened ion pseudopotential. The final potential is sensitive to the exact form of the core potential only when the separation between core and conduction electron states is small.

Having found the bare ion form factor, which theoretically contains all of the information necessary to describe the ion, we must subject this to screen-ing by the gas of conduction electrons. This is achieved in the linear approxi-mation, where the bare ion and screened ion form factors are related by a dielectric function which may take several forms, depending on how we include the exchange and correlation between the conduction electrons. In most cases the dielectric constant contains a logarithmic singularity which upon transformation of the screened ion form factor gives rise to oscillations in the real-space potential. We note that the dielectric constant itself may contain one or more parameters adjustable to experiment.

The total energy of the ion–electron system may now be found to the second order using perturbation theory. It consists of two contributions—the free-electron energy and the band structure energy. The former is independent of the ion positions (i.e., of the structure factor) depending only on the volume of the system, while the latter describes how the total energy varies with the ion positions at constant volume. It is from the band structure energy and the direct ion core–core interaction that we can derive a two-body potential, whose validity is restricted by reason of the basic assumptions to nonoverlapping ion cores and constant volume in the system. Quite evidently this potential is useful only in describing the interaction near the equilibrium separation in the crystal, as well as in a situation where the distribution of conduction electrons in the region of an ion resembles that in the case of a perfect crystal.

Although the development of pseudopotential theory represents a considerable advance in the theory of metals, its application is rather limited. A reliable and accurate first-principle calculation of the pseudopotential is difficult. In addition, the description of any phenomenon using pseudopotential theory hinges on the validity of the second-order perturbation treatment for the phenomenon in question. In principle one could define a pseudopotential which would give correct experimental values of any atomic property provided that the perturbation expansion is carried to sufficiently high order. However, a potential which correctly describes one property in the second-order perturbation approximation may not necessarily be reliable in the context of a different property. Any quest for an "accurate" pseudopotential should take this fact into account, although it must be admitted that the second-order perturbation approximation has had a good measure of success in explaining various phenomena.

For further exploration in depth of the topic of pseudopotentials the reader is referred to some comprehensive recent reviews of the subject by Heine and collaborators [21–23].

REFERENCES

1. N. F. Mott and H. Jones, "Theory of the Properties of Metals and Alloys." Oxford Univ. Press (Clarendon), London and New York, 1936.
2. J. M. Ziman, *Advan. Phys.* **13**, 89 (1963).
3. W. A. Harrison, "Pseudopotentials in the Theory of Metals." Benjamin, New York, 1966.
4. V. Heine and I. V. Abarenkov, *Phil. Mag.* **9**, 451 (1964).
5. A. O. E. Animalu and V. Heine, *Phil. Mag.* **12**, 1249 (1965).

6. R. W. Shaw and W. A. Harrison, *Phys. Rev.* **163**, 604 (1967).
7. R. W. Shaw, *Phys. Rev.* **174**, 769 (1968).
8. N. W. Aschcroft, *Phys. Lett.* **23**, 48 (1966).
9. P. S. Ho, *Phys. Rev.* **169**, 523 (1968).
10. J. Bardeen, *Phys. Rev.* **52**, 688 (1937).
11. J. Lindhard, *Kgl. Dansk. Vid. Selsk. Mat.-Fys. Medd.* **28**, No. 8 (1954).
12. L. J. Sham and J. M. Ziman, *Solid State Phys.* **15**, 221 (1963).
13. J. Friedel, *Phil. Mag.* **43**, 153 (1952).
14. J. Hubbard, *Proc. Roy. Soc.* **A243**, 336 (1958).
15. W. Geldart and S. H. Vosko, *Canad. J. Phys.* **44**, 2137 (1966).
16. L. J. Sham, *Proc. Roy. Soc.* **A283**, 33 (1965).
16a. R. W. Shaw and R. Pynn, *J. Phys. C* **2**, 2071 (1969).
17. K. S. Singwi, M. P. Tosi, R. H. Land, and A. Sjolander, *Phys. Rev.* **176**, 589 (1968); *Phys. Rev. B* **1**, 1044 (1970).
18. A. I. Gubanov and V. K. Nikulin, *Phys. Status Solidi* **17**, 815 (1966).
19. V. K. Nikulin and M. B. Trzhaskovskaya, *Phys. Status Solidi* **28**, 801 (1968).
20. W. A. Harrison, *Phys. Rev.* **181**, 1036 (1969).
21. V. Heine, *Solid State Phys.* **24**, 1 (1971).
22. M. H. Cohen and V. Heine, *Solid State Phys.* **24**, 38 (1971).
23. V. Heine and D. Weaire, *Solid State Phys.* **24**, 250 (1971).

CHAPTER VI

PAIR POTENTIALS BASED ON
PSEUDOPOTENTIAL THEORY

6.1 INTRODUCTION

In Section 5.9 we described how it is possible to write the pair inter-action potential as the sum of a direct interaction between ion cores and an indirect interaction via the intervening electron screening charge cloud. A necessary condition was that the conduction states should be well separated from the core states in energy, and that the ion cores should be small enough for the core states in a metal to be considered identical to those in the isolated ion. The indirect interaction could be calculated by assuming a reasonable potential to simulate the ion core, linearly screened by a gas of conduction electrons, allowing for exchange and correlation. The calculation could be performed from first principles or a model potential could be used, including parameters adjusted to experimental data. The latter is a good deal simpler in concept and often more reliable.

Basic steps leading to the two-body interaction were described in Chapter V and the final equation for the total (direct plus indirect) pair potential is

$$V(r) = Z^2 e^2/r + (\Omega_0/\pi^2 r) \int_0^\infty G(q) q \sin qr \, dq \qquad (6.1)$$

where

$$G(q) = -(\Omega_0 q^2/8\pi e^2) |w_b(q)|^2 \{1 - 1/\varepsilon(q)\} \qquad (6.2)$$

We note that Z is the effective charge number of the ion core (the nucleus of the pseudo-atom) rather than of the atom itself. This equation reveals quite clearly that the whole problem of establishing the interatomic pair potential of the system hinges on the determination of the energy-wavenumber characteristic $G(q)$, which as we recall comprises the bare ion form factor and the electronic screening contribution. Unless $G(q)$ is quite unrealistically simple, the integration must be performed numerically, but generally convergence with increasing q is rapid and no great difficulty is encountered. When the dielectric function $\varepsilon(q)$ contained in $G(q)$ possesses a logarithmic singularity at $q = 2k_F$ particular care must be taken to select small increments for numerical integration in this region in order not to neglect the effects of this singularity.

Most of the calculated pair potentials have unsurprisingly concerned the alkali metals as being those which best satisfy the conditions for the pseudo-potential formulation. However, the calculation has been extended by some workers to polyvalent metals such as Zn, Mg, and Al. We shall consider first of all potentials derived theoretically from first principles, in order to see whether they measure up to the test of comparison with experiment. Depending on what experimental quantities are involved, the comparison may take place either in wavenumber space or in the real space potential.

6.2   FIRST-PRINCIPLE DERIVATIONS OF THE INTERATOMIC POTENTIAL

As mentioned in Section 5.5, Harrison has calculated from pseudo-potential theory the OPW form factors $\langle k + q | w(r) | k \rangle$ for a number of metals. The method, which includes somewhat complicated mathematics, is not reproduced here. For our purposes it suffices to comment that the calculations are based on nonlocal pseudopotentials involving Hartree–Fock (HF) functions for the ions, linearly screened by a dielectric function excluding correlation and exchange. The Harrison OPW form factors were reproduced in Fig. 5.1. The real-space pair potentials resulting from these form

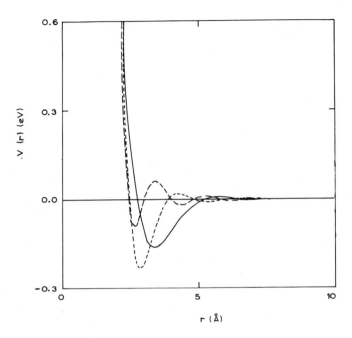

Fig. 6.1. Pair potentials of Pick [1] for Na (———), Mg (- - - -), and Al(– – –).

factors were calculated by Pick [1], and those for Na, Mg, and Al are illustrated in Fig. 6.1. Unfortunately, attempts to calculate the vibration spectrum for these metals [1, 2] using the theoretically derived $G(q)$ resulted in curves which were at variance with the experimental neutron scattering data where these were available. Harrison noted that the arbitrary addition of an energy factor to the pseudopotential for Al increased enormously the agreement between calculated and experimental vibration spectra, and recommended that the revised phenomenological form factor be used in calculation of atomic properties. The errors in the theoretical pseudopotential might, he observed, be due to mathematical approximations inevitable in the calculation, including the neglect of exchange and correlation in conduction electron interactions, or it might stem from the use of perturbation theory.

It is doubtless very satisfying to calculate the pseudopotential and through it the interionic potential directly from first principles without recourse to any experimental information other than the most basic atomic constants. However, the ability of such a potential to simulate nature as demonstrated by experiments leaves much to be desired. In view of the complication of the theoretical treatment and the present state of knowledge concerning the many-body problem, a simpler and more phenomenological type of pseudopotential would be rather more reliable for calculation of atomic properties, provided of course that some experimental information goes into its determination. We shall now proceed to consider some such empirical potentials.

6.3   THE HEINE–ABARENKOV–ANIMALU   (HAA) MODEL POTENTIAL

In formal pseudopotential theory the pseudopotential and pseudowave function are nonunique, while being of course complementary—that is, when one is defined the other follows. To reduce the problem to its most simple form, both the potential and the wavefunction should be as smooth as possible, to obtain rapid convergence in numerical calculations. Abarenkov and Heine [3] noted that if a smoothly varying pseudopotential was chosen a rapidly oscillating wavefunction ensued, and vice versa. This was a consequence of expanding the function as a complete set of vector waves, leading to an overcomplete vector space. They found that it was impracticable to search for a compromise by minimizing the oscillatory nature of both potential and wave function simultaneously, and suggested that a solution might be sought by calculating the pseudopotential based on a given smooth pseudowave function, then smoothing the potential by approximation. However, if one reconciles oneself to this form of treatment, it does not represent any further approximation to begin with a simple form of smoothed-out pseudopotential based on suitable experimental data. Such a potential was in fact suggested for the ion core in an earlier calculation of cohesive properties of metals by Cohen [4].

Abarenkov and Heine required their potential to have the same energy eigenvalues as the real ion potential. This meant that it had to be a summation over the angular momentum quantum number l, since different l electrons were orthogonal to different core functions. There was in addition a dependence on the energy level or principal quantum number. These factors led to a nonlocal potential involving a projection operator for the valence functions. Various forms were tried for their basic ion pseudopotential, the one finally chosen being that which had the weakest energy dependence for a fixed l. This was a sum of potentials each of which was a constant $A_l(E)$ for a given value of l. The constant was the result of an interpolation between the values found for different energy levels for a fixed l in the free atom or ion. It was multiplied by a projection operator P_l which picked out the component of the valence function with that value of l. Thus the model potential was of the form

$$w_b(r < R_m) = -\sum_l A_l(E)P_l$$

$$w_b(r > R_m) = -Ze^2/r$$

(6.3)

The Cohen ion core potential suggested earlier was the same as the above for $l = 0$, so that inside the ion core it was constant $(-V_0)$. Then the V_0 and ionic radius R_m were fitted to the lowest s- and p-levels determined from atomic spectra. We shall see in later sections how such a potential can be both easy to treat mathematically and extremely useful.

The HAA model potential of Eq. (6.3) is perhaps the empirical potential which follows most closely the spirit of pseudopotential theory, since it is both energy dependent and an operator. The basic assumption is the equivalence of the metallic and free ions. The ion is assumed to have an effective radius R_m outside which the Coulomb attraction prevails. Inside R_m the potential is constrained to simulate from the point of view of a conduction electron the energy eigenstates of the free ion. This is similar in general approach to the quantum defect method [5]. The $A_l(E)$ are fitted to the spectroscopically measured energy levels of the ion plus one electron (which is just the free atom in the case of the monovalent metals). The value of R_m is chosen between the core radius and that of the Wigner–Seitz cell, governed in subsequent work by the condition that $w_b(r)$ should be as continuous as possible at $r = R_m$. For values of l greater than 2 Heine and Abarenkov assumed that $A_1(E) = A_2(E)$, so that

$$w_b(r < R_m) = -A_2 + (A_0 - A_2)P_0 - (A_1 - A_2)P_1 \qquad (6.4)$$

since $\sum P_l = 1$.

If we were to compare the true ion potential with the model potential, we should find that they are exactly the same outside the radius R_m and possess the same radial derivative at $r = R_m$. Inside R_m the real potential is much stronger and more rapidly varying in character than the model, but this does not concern a conduction electron which does not penetrate the core.

Animalu and Heine [6] used the model potential to calculate the matrix elements between states on the Fermi surface, equivalent to the OPW form factors of Harrison. These form factors were illustrated in Fig. 5-2.

The discontinuity in the bare ion potential leads to oscillations in the form factor at large values of q, the shape and position of these depending markedly on the choice of the ion radius R_m. The question was, were these oscillations real or simply a feature of the model potential? Animalu and Heine eliminated the artificial part of the oscillation by making the discontinuity in $w_b(r)$ as small as possible. This corresponded to choosing the value of R_m such that the lowest energy level parameter A_0 was as close as possible to Ze^2/R_m. They also introduced a damping factor $D(q)$ which multiplied the screened ion form factor $w_s(q)$ and decreased rapidly after $q = 4k_F$:

$$D(q) = \exp\{-0.03(q/2k_F)^4\} \qquad (6.5)$$

This is quite an artificial factor designed to reduce the amplitude of the oscillating part of $w_s(q)$ but not to affect the first positive maximum of the function, which is a real feature of the band structure. Animalu and Heine showed that for a simplified local potential one could choose R_m within 10 to 20% of Ze^2/A_0 without significantly changing the phase or amplitude of the oscillations, but outside this limit the resulting form factors varied rather arbitrarily in amplitude and phase. They concluded that because of this latitude in the choice of R_m the oscillations might have physical significance but that in most applications they were sufficiently small to neglect. To calculate their screened ion form factors they chose R_m in accordance with the value of A_0. Tables of these form factors with and without the damping factor $D(q)$ are stated by Animalu and Heine [6] to be available on request from these authors. They have also been published, without $D(q)$ included, in the book by Harrison [2], although apparently in the case of the form factor for cadmium an error appeared in these tables which has since been rectified.

A comparison of the HAA form factors at the first few reciprocal lattice vectors with experimental values determined from Fermi surface studies in metals revealed good quantitative agreement. These form factors are certainly more realistic than those calculated by Harrison from the full pseudopotential theory, simply through the inclusion of some experimental data and the more empirical approach. They may be inverted to obtain a pairwise interaction between the ions in a similar fashion to that outlined in Chapter V.

This has in fact been carried out for the alkali metals by Shyu and Gaspari [7], who used a Hubbard modification of the Hartree dielectric function (see Section 5.6) similar to that given by Eq. (5.44) to obtain the function $G(q)$ for the inversion procedure. The resulting pair interaction potential is shown in Fig. 6.2 for Li. The long-range oscillations in the real-space potential are only significant in this model for Li, decaying very rapidly after about the fifth neighbor in the lattice for the other four metals of the series. This is a consequence of an abnormally large value of $w_s(q)$ in the HAA model for Li, since the amplitude of the oscillations depends on the square of the form factor. Using the pair potential obtained in this way Shyu and Gaspari calculated the tangential and radial force constants K_t and K_r at the nearest-neighbor sites in the lattice:

$$K_t = (1/r)(dV/dr)$$
$$K_r = d^2V/dr^2 \tag{6.5a}$$

They then analysed these in terms of Born–Von Karman force constants for comparison with those obtained to fifth nearest-neighbors by fitting the experimental phonon dispersion curves for Na and K. The result was a considerable degree of disagreement. Animalu et al. [8] have reproduced

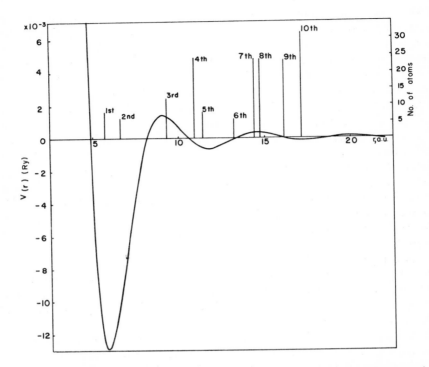

Fig. 6.2. Pair potential for Li derived by Shyu and Gaspari from the HAA model potential. Positions and numbers of neighbors are marked on the figure. (After Shyu and Gaspari [7].)

fairly satisfactorily the phonon dispersion curves by summing the force constants calculated directly from the HAA model potential out to fifth-neighbor sites. So some further investigation of the reasons for the Shyu–Gaspari disagreement seems to be necessary since the basic potential is the same. Motivated by the lack of agreement for the force constants, Shyu and Gaspari proceeded to calculate the elastic constants c_{11}, c_{12} and c_{44} through the medium of the secular equation, and this time found reasonable accord except in the case of Li.

An attempt by Torrens and Gerl [9] to reproduce the Shyu–Gaspari potentials by Fourier transformation of identical HAA screened ion form factors resulted in a certain amount of disagreement in the long-range part of the two sets of pair interactions (see Fig. 6.3). We exercised extreme care in the region of $q = 2k_F$ which gives rise to the large-r oscillations. In addition, our computer program reproduced accurately the Pick pair potentials and the Ho form of $V(r)$ (see Section 6.6).

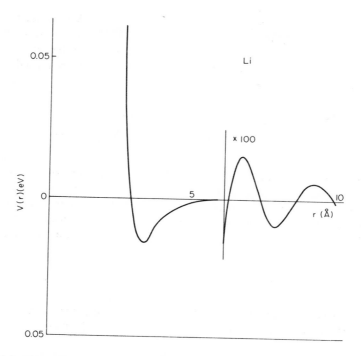

Fig. 6.3. Pair potential for Li derived from the HAA model potential. (After Torrens and Gerl [9].)

We mentioned briefly in Section 5.5 the more elaborate optimized model potential developed by Shaw. When this potential screened in the Hartree approximation was inverted to obtain the pair interaction, the first minimum near the nearest-neighbor position was missing [10]. For larger r the Shaw and HAA pair potentials are in reasonably good agreement although in the former case the oscillations die away more rapidly. Thus it seems that the $V(r)$ calculated from the optimized model potential using Hartree screening is not representative. However, Shaw makes the valid point that a purely local bare ion potential invariably leads to a first minimum in $V(r)$, so that the existence of such a minimum does not justify the model potential. The minimum does occur if electron-electron interactions are included (see Section 6.10). In fact, the energy-wavenumber characteristic in the optimized model form was found by Shaw and Pynn [10, 11] to be more accurate than the HAA model in predicting a number of atomic properties.

The already simplified approach of the HAA work inevitably leads one to wonder whether perhaps further simplification is possible, facilitating the mathematical analysis without losing much physical significance. After all, the core radius is considered a variable in the above work in order to smooth the form factors at large q. We shall go on in Section 6.5 to describe some further simplifications, but first we note briefly how the dielectric screening theory may be modified to apply to nonlocal bare ion potentials such as the HAA form.

6.4 THE SCREENING OF NONLOCAL POTENTIALS

Animalu [12] pointed out that the linear screening approximation leading to a dielectric constant $\varepsilon(q)$ which is a function of q only is rigorously only applicable to the screening of a local potential. Further complication is introduced when the potential contains nonlocal k-dependent terms, such as is the case for the HAA bare ion form. The unscreened nonlocal potential has a form factor which may be written

$$\langle \mathbf{k} + \mathbf{q} | w_b | \mathbf{k} \rangle = w_b(q) + \langle \mathbf{k} + \mathbf{q} | F | \mathbf{k} \rangle \qquad (6.6)$$

where it has been separated into local and nonlocal terms. The screening of the local part proceeds as before in the linear approximation:

$$w_s(q) = w_b(q)/\varepsilon(q) \qquad (6.7)$$

Making use of some work by Sham and Ziman [13] on screening of self-consistent potentials, Animalu derived an expression for the nonlocal screened form factor, which he wrote as follows:

$$\langle \mathbf{k} + \mathbf{q} | w_s | \mathbf{k} \rangle = w_s(q) + \langle \mathbf{k} + \mathbf{q} | F | \mathbf{k} \rangle + I(F) \qquad (6.8)$$

Here $I(F)$ is an integral expression involving $F(k)$ and the corresponding energy eigenvalues. It represents the screening field associated with the nonlocal part of the potential. The details of this development are beyond the scope of this book, and may be obtained by referring to the original articles.

Application of this refinement to the HAA model potential showed that the adjustment of the screened ion form factor was significant only in the region of $q = k_F$. Calculation of the resistivities of the alkali metals revealed better agreement with experiment when nonlocal screening was taken into account. But in general the differences between form factors calculated with and without nonlocal screening are not very great. When we go on to consider purely local bare ion potentials in the next few sections the simple q-dependent linear screening is all that we shall require.

6.5 THE ASHCROFT MODEL POTENTIAL APPLIED TO SIMPLE METAL PAIR INTERACTIONS

Ashcroft [14] suggested that a modified form of the model potential which omitted the energy dependence of the part inside the ion core would not lose physical significance for problems involving scattering at the Fermi surface. Since the pseudopotential formulation effectively almost cancels the large potential inside the ion core, the starting hypothesis of Ashcroft is that the cancelation is complete. Thus

$$w_b(r < R_m) = 0$$
$$w_b(r > R_m) = -(Ze^2/r) \tag{6.9}$$

We can see that this resembles the model potential suggested by Cohen [4] with $V_0 = 0$. In this form $w_b(r)$ is a purely local potential depending only on r. It was assumed that any incomplete cancelation could be taken into account by the variation in the choice of R_m, which was in the Ashcroft model found either from liquid metal resistivity or Fermi surface data.

The values of R_m used by Ashcroft are compared with the HAA values in Table 6.1. We note that the former are considerably lower than the latter

TABLE 6.1

COMPARISON OF ASHCROFT AND HAA BARE ION RADII (R_m) FOR THE ALKALI METALS

	R_m (a.u.)				
	Li	Na	K	Rb	Cs
Ashcroft	2.00	1.66	2.13	2.13	2.16
HAA	2.84	3.40	4.16	4.35	4.92

in all cases. This is not surprising when we take into account the restriction imposed on the system by requiring complete cancellation. R_m is an effective core radius and is only loosely related to the true ionic radius. Indeed the Ashcroft core radius is not far removed from the ionic radius as deduced from ionic crystal data, while the HAA core is about twice as large. Using the smaller core radius decreases the relative importance of the core terms in the potential, which serves to improve the validity of the Ashcroft model.

After inversion, the Ashcroft form factor is given by Eq. (5.25). Linear screening is again assumed in the same approximation as that used by Heine

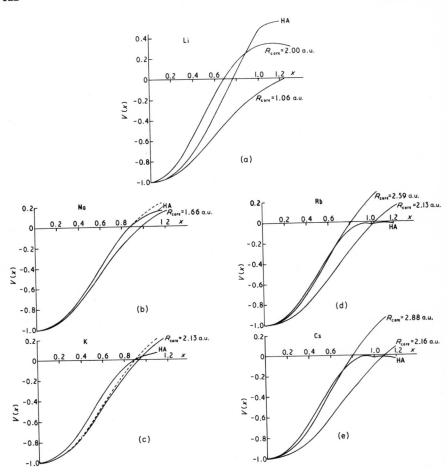

Fig. 6.4. Form factors of Ashcroft for the alkali metals with different values of the ion radius R_m (denoted in this figure by R_{core}). These are compared to Heine–Abarenkov (HA) form factors. In the case of Na and K the dotted curve corresponds to an Ashcroft form factor with R_m fitted to the first node of the HA form factor. (After Ashcroft [16].)

and Abarenkov (see Eq. (5.44)). Some form factors obtained by Ashcroft for the alkali metals are compared in Fig. 6.4 to the equivalent HAA form factors. The dotted curve illustrates that the Ashcroft model potential may be fitted with moderate success to the HAA form factors by choosing the value of R_m so that the two curves coincide where they cross the zero. If one places greater faith in the HAA method but would prefer a simpler and more analytically tractable formulation, this is a convenient way to approach the problem.

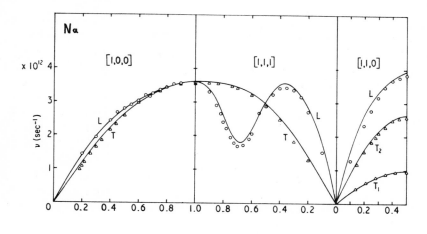

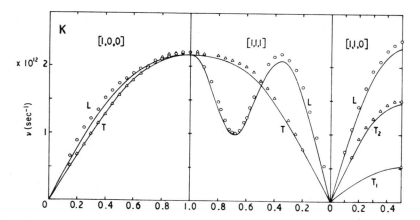

Fig. 6.5. Phonon dispersion curves of Na and K calculated using the Ashcroft potential taking into account 12 shells of near neighbors. These are compared with experimental phonon spectra of Cowley *et al.* [17]. The abscissae are plotted in units of $q[100]$, $q/2[110]$ and $q/3[111]$. (After Ashcroft [16].)

The dielectric exchange correction used by Ashcroft was that suggested by Geldart and Vosko [15] involving compressibilities of interacting and noninteracting electron gases (see Section 5.6). With this included in the screened ion form factors, Ashcroft calculated the phonon spectra for Na and K taking into account 12 shells of near neighbors and obtained good agreement with experimental curves [16] (see Fig. 6.5). Agreement with such things as liquid metal resistivity, ionic separation in the solid, compressibility and band-gaps was shown to depend rather sensitively in most instances on the experimental quantity used in the estimation of R_m, especially for Rb and Cs. This may be an indication of the necessity to take nonlocal potentials into consideration, since core-conduction band separation is small for these metals.

Shyu and Gaspari [18] carried out a determination of the interatomic pair potential for the Ashcroft form factor as they had for the HAA case, extending the calculation this time to 12 simple metals. Again it was a question of screening the bare ion by a modified Hubbard dielectric constant, finding the resulting function $G(q)$, and integrating according to Eq. (6.1). For those who wish to refer to the original reference, the $G(q)$ quoted by Shyu and Gaspari is slightly different from that which is defined in Eq. (6.2), by a factor of a constant multiplied by q^2. This has no bearing on the final result and is simply a matter of definition of terms.

TABLE 6.2

Shyu–Gaspari Matched Bare Ion Radii (R_m) for a Number of Metals, Compared with Ashcroft Values for the Alkalis

	R_m (a.u.)				
	Li	Na	K	Rb	Cs
Shyu and Gaspari	1.68	1.76	2.13	2.26	2.43
Ashcroft	2.00	1.66	2.13	2.13	2.16
	1.06				

	Be	Mg	Ca	Zn	Cd	Al	Ga
Shyu and Gaspari	1.06	1.36	1.91	1.11	1.25	1.13	1.12

Placing their confidence in the numerical values of the HAA form factor, Shyu and Gaspari determined the core radius R_m as outlined above, by matching the first node of the Ashcroft $w_b(q)$ to that of HAA. The values of R_m thus found were not greatly different to those of Ashcroft, as may be seen from Table 6.2. They observed that the value of R_m found in this way was quite insensitive to the exact point of matching to the HAA potential between $q = 0$ and the first node.

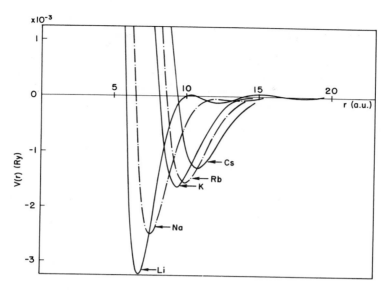

Fig. 6.6. Alkali metal potentials derived by Shyu and Gaspari from the Ashcroft model potential. (After Shyu and Gaspari [18]).

The form of the interionic potentials obtained thus by Shyu and Gaspari is illustrated in Fig. 6.6. We can make several qualitative points about these in comparison with their earlier potentials based directly on the HAA form factors (see Fig. 6.2). The oscillation at large r is much more evident in the polyvalent metals than in the alkalis. Once again Li proves an exception, undoubtedly due to the influence of the HAA form factor. The well-depths in the vicinity of the nearest-neighbor separation are in good agreement with the corresponding HAA potential well-depths, except for Li, where the HAA minimum is four times deeper. We note also that the oscillations occasionally occur at potential energies removed from the zero, chiefly for the polyvalent metals.

The potentials for the cubic metals were again tested by using them to obtain values for the binding energy, interatomic force constants and elastic constants. The binding energy was estimated as a function of the crystal structure at constant volume for FCC, BCC, and HCP structures. The results were inconclusive for the alkali metals, all three binding energies falling within the convergence of the summation. However, the correct crystal structure was predicted for Ca and Al. Better quantitative agreement with experiment than in their earlier work was evident in the interatomic force constant calculation for the alkali metals, although there were still some

numerical discrepancies. Such was also the case for Al. The experimental
elastic constants were on the other hand well reproduced for all the alkalis
except Li, but the agreement for Al was not so good. Although they did not
calculate the phonon dispersion curves using their version of the Ashcroft
form factor, Shyu and Gaspari extrapolated from their elastic constant computations to suggest that they should obtain reasonably good reproduction
of the experimental curves.

6.6 The Pair Potential of Ho for the Alkali Metals

We have already described briefly the bare ion form factor and exchange
screening method of Ho in the last chapter (see Sections 5.5 and 5.6). Essentially, his bare ion potential is of the type suggested by Cohen [4]:

$$w_b(r < R_m) = -V_0$$
$$w_b(r > R_m) = -Ze^2/r \tag{6.10}$$

which leads to the form factor of Eq. (5.27). The screening followed closely
the Hubbard approximation, the dielectric constant being given by

$$\varepsilon(q) = 1 + \frac{g_H(q)}{q^2}\left\{1 - \frac{g_H(q)}{2[q^2 + f(k_F)]}\right\}^{-1} \tag{6.11}$$

Ho included a parameter in this exchange term of the dielectric constant
through the function $f(k_F)$:

$$f(k_F) = \eta k_F^2 \tag{6.12}$$

This factor η combined with the V_0 and R_m to provide three adjustable
parameters for comparison with experimental results.

The experimental data chosen by Ho for determination of his model
potential parameters were the elastic constants c_{11}, c_{44}, and $c' = (c_{11} - c_{12})/2$.
Through the medium of the secular equation for phonon frequencies, the

TABLE 6.3

Potential Parameters of Ho
for 4 Alkali Metals

	Li	Na	K	Rb
V_0 (eV)	11.49	7.34	5.58	5.44
R_m (Å)	1.18	1.22	1.59	1.74
η	1.58	1.78	1.87	2.02

long-wave limits of these three elastic constants were expressed as functions of the screened ion potential and its first two derivatives summed over reciprocal lattice vectors. In this way Ho was able to choose the three parameters of the potential to fit the elastic constants to within 3 %. The V_0, R_m, and η for the first four alkali metals given in the paper by Ho are reproduced in Table 6.3.

Using the resulting model potential, Ho went on to calculate the phonon dispersion curves for the alkali metals, which resulted in excellent agreement with experiment in the known cases of Na and K (see Figs. 6.7 and 6.8). Finally, the two-body real-space potentials for the four metals are reproduced in Fig. 6.9. At the time of writing his paper, Ho did not have available the experimental elastic constants of Cs, so its potential could not be calculated by this method. As may be seen from the figure, these potentials are similar to those deduced from other model potentials, although the oscillations are much more rapidly decaying, the depth of the second minimum being less than one-hundredth of that of the first.

The Ho potential therefore takes into consideration the electron screening contribution during adjustment of its parameters, instead of fitting only the bare ion constants to experiment. Unfortunately the adjustment takes place at the long-wave limit (small q), and it is evident that the critical function $(1 - 1/\varepsilon(q))$ is rather insensitive to η for small q. So it would be desirable to obtain this constant in some other fashion. Ho has recently [20] redetermined his model potential parameters, this time using the compressibility of the free electron gas [15] to obtain η. The revised parameters are given in Table 6.4. This time Cs is included since its elastic constants are now available.

TABLE 6.4

REVISED POTENTIAL PARAMETERS OF HO
FOR THE ALKALI METALS

	Li	Na	K	Rb	Cs
V_0 (eV)	9.38	7.75	5.17	5.44	5.03
R_m (Å)	0.89	1.27	1.60	1.87	2.09
η	1.84	1.81	1.77	1.76	1.74

The new Ho potential might be considered more reliable in that it adjusts the screening term to experiment in a more direct fashion, but the difference between the two potentials is relatively small in view of the empirical nature of the basic model.

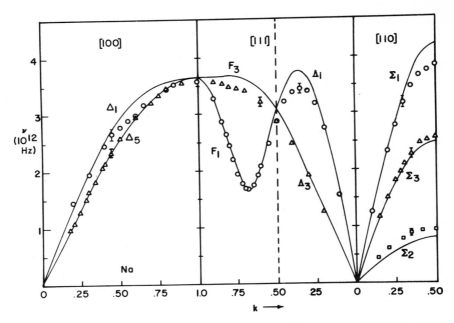

Fig. 6.7. Phonon dispersion curves calculated by Ho for Na, compared with the experimental curves of Cowley *et al.* [17]. (After Ho [19].)

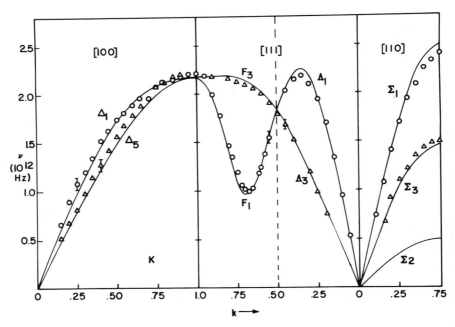

Fig. 6.8. Phonon dispersion curves calculated by Ho for K, compared with the experimental curves of Cowley *et al.* [17]. (After Ho [19].)

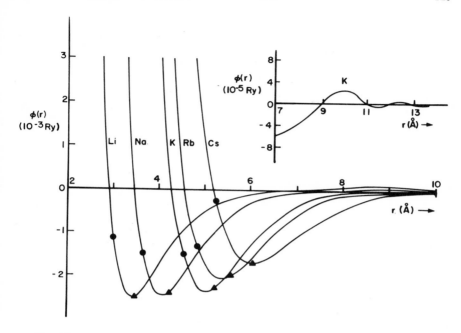

Fig. 6.9. Alkali metal pair potentials of Ho. The long-range part of the potential for K is magnified ×100 in the inset. (After Ho [19].)

6.7   PAIR POTENTIALS DETERMINED FROM PSEUDO-ATOM PHASE SHIFTS

In the spirit of pseudopotential theory the pseudo-atom is regarded as an entity in the metal, made up of the bare ion and its screening charge cloud. Then the normal methods of partial wave analysis may be applied to it. Let us very briefly review the partial wave method. The asymptotic partial wave solution of the Schrödinger equation for a spherically symmetric central potential is of the form

$$\psi_l = (1/r)\sin(kr - l(\pi/2) + \eta_l)P_l(\cos\theta) \qquad (6.13)$$

where P_l is the lth Legendre polynomial and η_l the corresponding phase shift. A very useful feature of phase shift analysis is the Friedel sum rule [21]:

$$Z = (2/\pi)\sum_l (2l + 1)\eta_l(k_F) \qquad (6.14)$$

The symbol η_l used here for the phase shifts should not be confused with the constant η of the previous section.

This rule-of-thumb arises from a comparison of the number of electrons required to fill electron states up to $k = k_F$ with and without the potential source present, Z being the charge of the source. The factor $(2l + 1)$ is the number of states of a given l, and the 2 comes from the electron spin states. The utility of the sum rule lies in its ability to limit the possible phase shifts of a given potential and thus to define the potential. Suppose we have a certain central potential containing an unknown parameter. The partial wave solution of the Schrödinger equation with this potential may be calculated and the parameter adjusted until the resulting phase shifts satisfy the Friedel sum rule.

The problem of scattering from such a potential may also be subjected to the partial wave approach. Matching a plane wave $e^{\mathbf{k} \cdot \mathbf{r}}$ to a partial wave of the form (6.13) yields a differential scattering cross section given by

$$f(\theta) = (1/k) \sum_l (2l + 1) e^{i\eta_l} \sin \eta_l P_l(\cos \theta) \tag{6.15}$$

This scattering amplitude may be used together with the sum rule to estimate for example the scattering of electrons associated with a heterovalent impurity in a metal, assuming a given form of impurity potential. Hence the resistivity associated with the impurity may be calculated.

Meyer, Nester, and Young [22] have calculated the partial wave phase shifts associated with a pseudo-atom in a metal. For the pseudo-atom the free atom Hartree–Fock–Slater core functions given by Hermann and Skillman [23] were used in conjunction with a valence potential of the bare ion type and a core-orthogonalization term. The bare ion potential contained a constant adjustable parameter related to the cutoff radius:

$$\begin{aligned} w_{\mathrm{b}}(r < 1/\varepsilon) &= \varepsilon \\ w_{\mathrm{b}}(r > 1/\varepsilon) &= -(Ze^2/r) \end{aligned} \tag{6.16}$$

The core-orthogonalization potential was chosen in the semilocal approximation, by assuming that inside the core cancelation with the HF potential was good. Thus an electron of a certain l experienced a simple r-dependent potential which would of course vary with different l. This is the same approach as that of Heine and Abarenkov (see Section 6.3). The only further approximation made by Meyer, Nester, and Young (MNY), based on this cancelation inside the core, was that the radial part of the l-dependent electron wave function in the core varied as r^l. Then the radial partial wave equation could be set up and the phase-shifts for different l calculated at various values of k. The value of the parameter ε of Eq. (6.16) for different values of k_{F} was established by adjusting it until the phase shifts thus determined satisfied the Friedel sum rule.

Meyer, Nester, and Young tabulated the values of ε and the first four phase shifts η_0, η_1, η_2, and η_3 found by this method for the alkali and noble metals and listed several limitations of their model. First, the Slater approximation to the HF field was considered rather poor for Li. Second, the r^l

approximation for the wavefunction inside the core is good for the s- and p-waves but loses validity for d-waves in the outer part of the core (η_2). The metals to which this objection specifically applies are Rb and Cs, since for other alkali metals the bound d-state is absent, and for the noble metals the phase shifts η_2 are small. Finally, the phase shift determination is limited to an intermediate range of k_F, since at low k_F or low electron density the tendency will be to form a new bound state thereby changing fundamentally the electron configuration, and at high k_F one of the η_l will tend to dominate the Friedel sum.

The method outlined above had a certain amount of success in accounting for the transport properties particularly of the alkali metals, especially when one considers the very minimal amount of experimental data introduced in the phase shift determination. Meyer, Young, and Dickey [24] (MYD) subsequently proceeded to ask whether the interionic forces in a metal could be described within the framework of this theory. The form of interatomic potential which could be most simply obtained from the electron–ion phase shifts was the asymptotic potential

$$V(r)_{\text{asymp}} = A \cos(2k_F r + \phi)/r^3 \qquad (6.17)$$

Comparing this with Eq. (5.63) we see that the constant A contains the square of the bare ion potential evaluated at $k = 2k_F$. The charge cloud polarization induced by an ion may be expressed in terms of the phase shifts by the equation

$$\Delta n \sim (1/2\pi^2 r^3) \sum_l (2l + 1)(-1)^{l+1} \sin \eta_l(k_F) \cos[2k_F r + \eta_l(k_F)] \qquad (6.18)$$

In MYD terms Δn was written in the form

$$\Delta n \sim a \cos(2k_F r + \phi)/r^3 \qquad (6.19)$$

where (in atomic units)

$$a \cos \phi = (1/2\pi^2) \sum_l (2l + 1)(-1)^{l+1} \sin \eta_l \cos \eta_l$$

$$a \sin \phi = (1/2\pi^2) \sum_l (2l + 1)(-1)^{l+1} \sin^2 \eta_l \qquad (6.20)$$

So when a second bare ion is introduced in order to find the pair interaction, the comparison of the resulting potential with the asymptotic $V(r)$ of Eq. (6.17) shows that

$$A = a w_b(2k_F) \qquad (6.21)$$

The MYD bare ion potential was simply given by

$$w_b(2k_F) = -(\pi/k_F^2)\rho(2k_F) \qquad (6.22)$$

where $\rho(2k_F)$ is the total charge density evaluated at $k = 2k_F$, following

Poisson's equation. It has also been tabulated in Meyer *et al.* [22], being derived from the wavefunctions used in the calculation.

Knowing the appropriate phase shifts η_l and $\rho(2k_F)$ from the MNY work it was possible to calculate the amplitude and phase A and ϕ of the asymptotic form of $V(r)$ by way of Eqs. (6.19)–(6.22), thus establishing an interatomic potential valid except at small values of r. This potential has an oscillatory form similar to the large-r part of the Pick pair potential (see Fig. 6.10).

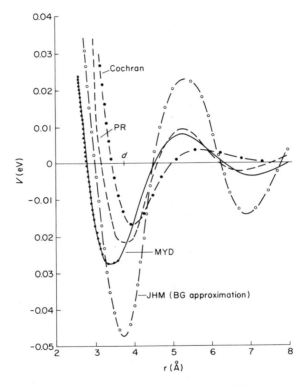

Fig. 6.10. Interionic potentials for Na obtained by MYD, compared with that of Cochran [29] from phonon spectra (see Sections 6–9), and the liquid metal potentials of JHM [26] and PR [27] (see Chapter IX). (After Meyer *et al.* [24].)

Meyer, Young, and Dickey pointed out in connection with their asymptotic form that if one compares Eqs. (6.20) and (6.15) there follows the simple relation

$$ae^{i\phi} = -(k_F/2\pi^2)f(\pi) \tag{6.23}$$

Here $f(\pi)$ is the scattering amplitude for back-scattering. This facilitates comparison of the potential with those derived from such experimental data as the resistivity of the liquid metal, which is largely dependent on back-scattering.

The question of validity of the interatomic potential at the nearest-neighbor distance in the crystal is of importance in application to atomic problems. Meyer, Young, and Dickey suggest that it is reasonable to assume that their potential is a good approximation at the nearest-neighbor separation since in the case of $l = 2$ or larger, when the asymptotic form would be questionable at this distance, the contribution of the corresponding η_l to the Friedel sum is negligible. However, it should be remembered that this is only an asymptotic form of the potential.

There are two objections to the approach of Meyer and co-workers which should be mentioned. One is connected with this asymptotic nature of the ionic interaction. The method might be a good one for two isolated impurities in an electron gas at large distance but may break down in the case of a relatively close-packed assembly of pseudo-atoms. The second criticism concerns the phase-shift ϕ of Eq. (6.17), as compared with that of Eq. (6.19). It is only in the approximation of a free electron gas that a unique phase shift may be assumed for two interacting ions. In reality both ions are screened by an electron charge cloud and consequently it is necessary to consider the interaction of the two screened charges, which involves two separate phase shifts ϕ. In the asymptotic form then the phase-shift of the interionic potential in Eq. (6.17) should be 2ϕ for two identical ions.

6.8 Impurity Potentials and Alloys

When an impurity atom is present in infinitely dilute solution in a matrix of metal atoms, the calculation of the pair potential between the impurity and matrix atoms by pseudopotential methods requires a careful consideration of the physical situation created by the substitution process. Let us first look at the most elementary type of impurity, namely one which is homovalent with the matrix. The simplest hypothesis is that the core wavefunctions of the impurity are unaltered in passing from the free atom to the metallic matrix (which is what we do in pseudopotential theory when we put the ion into its own matrix). Occasionally, as for example in the case of noble metal impurities, this assumption can be more valid for the ion as an impurity than as an atom in its own pure metal, when the large d-states overlap and become distorted. Then we might simply assume that the electron screening is unaltered by the substitution and divide the bare ion form factor by the dielectric constant of the matrix. Calling our matrix and impurity atoms respectively 1 and 2, we have for the screened ion form factors

$$w_{s1}(q) = w_{b1}(q)/\varepsilon_1(q)$$
$$w_{s2}(q) = w_{b2}(q)/\varepsilon_1(q) \tag{6.24}$$

Following these assumptions we may proceed to calculate the impurity-matrix pair potential in the usual way. Such a procedure was followed in a calculation by Torrens and Gerl [9] of impurity alkali–alkali metal potentials to be used in point defect atomistic energy calculations, the pure metal form factors being those of Ho [19] (see Fig. 6.11). It should be noted that there is an additional complication here, since the parameters contained in the bare ion form factor of Ho relate the ion potential to the ion in its own metal matrix rather than to the free ion. This is a result of the parameters being obtained from the pure metal elastic constants. Consequently, Eq. (6.24) assumes that the bare ion potential does not change in passing from its own metal to another.

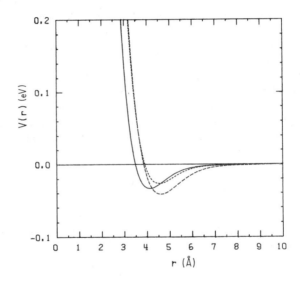

Fig. 6.11. Pair potentials for homovalent alkali impurities in alkali metals: (——) Na–Na; (– – – –) K in Na; (– – –) Na in K.

The only modification of this elementary treatment required for a heterovalent impurity would be the addition of a bare ion form factor of the nature of that given in Eq. (5.28) for a point charge, so that the screened ion form factor becomes

$$w_{s2}(q) = (1/\varepsilon_1(q)\{w_{b2}(q) - (4\pi \, \delta Z \, e^2/\Omega_0 q^2)\} \tag{6.25}$$

Here δZ is the difference in valence between the impurity and matrix.

The approximations involved in an impurity potential such as that described above are evident. First, when a free ion is inserted in a metal as an impurity, its environment is generally altered both from its free atom and pure metal surroundings. This leads to a change in the core shift Δ, i.e., the energy zero of the crystal relative to that in the situation used to define its form factor, be it free or metallic ion. There is also a change in the value of k_F. Where the form factor is obtained by parameter adjustment in the metal itself the pure metal core shift is zero, but when the ion becomes an impurity in another metal, Δ becomes nonzero. The shift in k_F will have its greatest effect on the dielectric constant, which is a function of q/k_F. It will also be important in calculating the form factor matrix elements in the case of a nonlocal pseudopotential. We must also bear in mind the change in atomic volume Ω_0, and related to this the size effect. If our impurity has a large ion core compared to the available space in the host lattice, the small-core approximation may break down with the introduction of closed shell repulsive forces. Finally, there may be a small change in the orthogonalization charge, but this is liable to be insignificant compared to the other corrections.

A simplified impurity representation was used by Harrison [2] to calculate resistivities of dilute alloys. A point ion model with delta-function strength V_0 plus a valence difference term constituted the bare ion form factor. The values of V_0 for both solute and solvent atoms were established by matching to the form factors obtained from *a priori* pseudopotential theory of the pure metals. The k_F, hence $\varepsilon(q)$ and Ω_0, were assumed to be those of the solvent. Screening was accounted for in the simple Hartree approximation. As might be expected from the extent of the approximation, agreement of calculated and experimental resistivities was limited to a few alloys where it probably owed more to chance than to theory. Harrison attempted to turn the problem around for solvents where his full theory form factor was unavailable, and adjust the delta-function strengths to obtain best agreement of computed and measured resistivities. In this way the form factor thus computed could be compared with that obtained by fitting the point–ion model to the HAA form factors more extensively available. Again, however, very little agreement was evident. Although the corrections mentioned in the last paragraph could be influential in deciding the accurate form factor for use in such a resistivity calculation, it is very probable that the use of the point–ion model and Hartree screening (which neglects electron–electron interactions) also produces major errors. It would be necessary in a more comprehensive study to estimate the relative importance of these different approximations and corrections.

Further development of this work on dilute alloys has been performed by Gupta [28], who based his form factor on the Harrison model for comparison purposes, but incorporated the various corrections such as shifts in k_F, Δ, and orthogonalization charge, and estimated the effect of each correction on the calculated resistivity. The residual resistivity was computed for

Mg, Zn, and Ca as impurities in Al, and for Al and Li in Mg. The corrections were introduced progressively—first of all the change in k_F and Δ, then non-local screening and exchange (see Sections 5.6 and 6.4), third, changes in the effective mass due to core orthogonalization. Agreement with experiment improved as more corrections were included although the major correction by far was the adjustment of Δ and k_F.

Gupta has calculated by this method the form factors and resulting resistivities of homovalent alkali impurities in alkali metals and made available these numerical form factors. However, since these are presently calculated only for q less than $2k_F$, we are unable to compute the impurity-matrix pair interaction corresponding to the Gupta form factors.

Since the model potential has in general been more successful in its application to atomic and electronic problems than the theoretically derived pseudopotential, it would be interesting to see whether a model potential modified for alloys along the lines described by Gupta would achieve a better correspondence with experiment.

Before leaving the present topic it would be well to emphasize that all the work described has been for infinitely dilute alloys, i.e., for effectively isolated impurities in the host lattice. If the concentration of impurities becomes significantly large the impurities will not be sufficiently well separated for their interaction to be neglected. Consequently, one might suppose that it will always be necessary in the case of an alloy AB to obtain three interatomic pair potentials V_{AA}, V_{BB}, and V_{AB}. It can be shown, however [2, 25], that if all the sites are equivalent in the crystal these three may be replaced by two potentials, a mean potential and a difference (or alloying) potential, which are more convenient for many studies.

Suppose that the alloy is made up of a number N of A and B type ions whose screened ion potentials are w_A and w_B, that the concentration of type B is c, and that the structure factors are $S(q)$, $S_A(q)$, and $S_B(q)$, where

$$S_A(q) = (1/N) \sum_i^{(A)} e^{i\mathbf{q}\cdot\mathbf{R}_i} \tag{6.26}$$

and

$$S(q) = S_A(q) + S_B(q) \tag{6.27}$$

The weighted mean of the bare ion pseudopotentials is

$$\bar{w} = (1 - c)w_A + cw_B \tag{6.28}$$

The total pseudopotential $W(q)$ is (see Eq. (5.13))

$$W(q) = \{(1 - c)w_A + cw_B\}S(q) + c(w_A - w_B)S_A(q) - (1 - c)(w_A - w_B)S_B(q) \tag{6.29}$$

$$W(q) = \bar{w}S(q) + cw_d S_A(q) - (1 - c)w_d S_B(q) \tag{6.30}$$

where w_d is the difference potential $(w_A - w_B)$.

We can now define functions $\bar{G}(q)$ and $G_d(q)$ by comparison with the energy-wavenumber characteristic introduced in Section 5.8, which correspond to the mean potential $\bar{w}$ and difference potential w_d respectively, assuming linear dielectric screening. Then the band structure energy of the above system is (cf. Eq. (5.52))

$$E_{bs} = \sum_g |S(q)|^2 \bar{G}(q) + \sum_q |cS_A(q) - (1 - c)S_B(q)|^2 G_d(q) \qquad (6.31)$$

The function $G_d(q)$ is the alloying energy-wavenumber characteristic. That no cross terms appear in the squaring of $W(q)$ is easily proved in the case of a Bravais lattice. If $\mathbf{q}$ is a reciprocal lattice vector then $e^{i\mathbf{q}\cdot\mathbf{r}_i} = 1$ for all $\mathbf{r}_i$ and the terms involving w_d cancel in Eq. (6.30). If $\mathbf{q}$ is not a reciprocal lattice, vector $S(q) = 0$ and the term in $\bar{w}$ vanishes. Thus if we sum over all $\mathbf{q}$ to obtain the band structure energy the resulting expression is that of Eq. (6.30). The same conclusion applies for more general structures [25].

To obtain pair potentials of the system we simply follow the development of Section 5.9, with the result as follows:

$$V(r) = \frac{(Z_A + Z_B)^2}{4r} + \frac{\Omega_0}{4\pi^3} \int \bar{G}(q) e^{i\mathbf{q}\cdot\mathbf{r}} d^3q \qquad (6.32)$$

$$V_d(r) = \frac{(Z_A - Z_B)^2}{4r} + \frac{\Omega_0}{4\pi^3} \int G_d(q) e^{i\mathbf{q}\cdot\mathbf{r}} d^3q \qquad (6.33)$$

Thus the ionic interactions may be expressed as the sum of a mean potential and a difference or alloying potential. The mean potential is just that of a solid which has the same crystal structure as the alloy and an ionic pseudo-potential $\bar{w}$. The difference potential is the one which determines the ordering of the alloy and is the more interesting of the two in atomic problems involving alloys.

Applying this type of alloy pseudopotential theory to binary alloys of Hg, Mg, and Cd, Inglesfield [25] had some success in predicting structures, ordering energies and solubility limits. These alloys are however judiciously chosen to avoid complications such as volume changes on solution or valence difference, which would affect the ionic pseudopotential. For more general classes of alloy one would have to consider changes in Fermi energy, electronic configuration and form factors of the ions when put into the alloy matrix, with corresponding increases in uncertainty and computational difficulties.

6.9 Derivation of the Pair Interaction from Phonon Spectra

We have noted at several points throughout the present chapter that the wavenumber space pseudopotential has been used to reproduce the phonon dispersion curves of the metal for the purpose of comparison with those obtained from neutron scattering techniques. Usually the potentials were defined in some other manner. For example the HAA model potentials relied on free ion spectra, the Ho potential was found from crystal elastic constants, and the Ashcroft form factor made use of liquid metal resistivity and Fermi surface data. There is no great difficulty in imagining a reversal of this procedure in order to determine the potential from the starting-point of the experimental phonon dispersion curves. It would be useful before describing such an analysis to sketch very briefly the connection between the pseudopotential and lattice vibration theory.

In a crystal lattice each atom will vibrate under the influence of the force-fields of neighboring atoms, which can in a simple model be averaged to a certain potential well and assumed to remain constant. In reality, since the neighboring atoms are also vibrating this field will tend to vary in such a way as to minimize the total energy, producing correlations between the oscillations of neighboring atoms. Thanks to the translational symmetry of the lattice this complicated problem reduces to solving a set of $3n$ simultaneous equations of motion of the form

$$m^{(n)}\omega^2(\mathbf{q})u_i^{(n)}(\mathbf{q}) = \sum_j M_{ij}^{(n)}(\mathbf{q})u_j^{(n)}(\mathbf{q}) \tag{6.34}$$

where there are n atoms per unit cell, $m^{(n)}$ is the mass of the nth atom, $u_i^{(n)}(\mathbf{q})$ its polarization vector in the ith coordinate direction, and $\omega(\mathbf{q})$ the frequency. $M_{ij}^{(n)}(\mathbf{q})$ is the force tensor involving second derivatives of the lattice potential with respect to the coordinates i and j. For a crystal with one atom per unit cell this reduces to three equations:

$$m\omega^2(\mathbf{q})u_i(\mathbf{q}) = \sum_j M_{ij}(\mathbf{q})u_j(\mathbf{q}) \tag{6.35}$$

The force tensor for a system of ions and electrons may be divided up into three contributions—one from the ion–ion closed shell repulsion, another from the direct Coulomb interaction between the ions, and a third from the screening by conduction electrons. The last two are grouped together as far as their potential is concerned in the concept of the pseudo-atom, where the ion with its conduction electron screening cloud is considered an entity in the solid. So it should be possible to group their contributions to the force tensor. This was the approach used by Cochran [29], who derived the term in the force tensor due to two sets of charge distributions $\rho_1(r)$ and $\rho_2(r)$ centered on the lattice points. The contribution is Coulombic in nature, given by

$$C_{ij}^{12}(\mathbf{q}) = \frac{4\pi}{\Omega_0} \sum_g \frac{(\mathbf{g} + \mathbf{q})_i(\mathbf{g} + \mathbf{q})_j}{|\mathbf{g} + \mathbf{q}|^2} \rho_1(\mathbf{g} + \mathbf{q})\rho_2{}^*(\mathbf{g} + \mathbf{q}) \tag{6.36}$$

where $\mathbf{g}$ is a reciprocal lattice vector and i, j denote components in the appropriate directions. If we write the two charge distributions as $\rho_b(r)$ and $\rho_e(r)$, the bare ion charge distribution and that due to the conduction electrons respectively, the result (6.36) applies for $E_{ij}(q)$, the screening contribution to M_{ij}. It now becomes a question of expressing these charge distributions in terms of the effective potential, which may be achieved through the pseudopotential formulation using Poisson's equation.

In our usual notation we relate the screened ion and bare ion form factors through the dielectric constant

$$w_s(q) = w_b(q)/\varepsilon(q) \tag{6.37}$$

We can then define by subtraction the potential due to the conduction electrons

$$w_e(q) = w_b(q)\{1/\varepsilon(q) - 1\} \tag{6.38}$$

Now using Poisson's equation

$$w_b(q) = -(4\pi e^2/\Omega_0 q^2)\rho_b(q) \tag{6.39}$$

so that

$$\rho_e(q) = \rho_b(q)[1/\varepsilon(q) - 1] \tag{6.40}$$

and

$$\rho_b(q)\rho_e{}^*(q) = -[\Omega_0 q^2/4\pi e^2]^2 |w_b(q)|^2[(\varepsilon(q) - 1)/\varepsilon(q)] \tag{6.41}$$

Comparison of Eqs. (6.41) and (6.2) relates the charge density product to the energy-wavenumber characteristic $G(q)$:

$$\rho_b(q)\rho_e{}^*(q) = (\Omega_0 q^2/2\pi e^2)G(q) \tag{6.42}$$

It follows that the screening contribution to these two charge distributions may be written, after Eq. (6.36),

$$E_{ij}(\mathbf{q}) = (4\pi/\Omega_0) \sum_Q (Q_i Q_j/|\mathbf{Q}|^2)\rho_b(Q)\rho_e(Q) \tag{6.43}$$

$$= (2q^2/e^2) \sum_Q (Q_i Q_j/|\mathbf{Q}|^2)G(Q) \tag{6.44}$$

where now

$$\mathbf{Q} = \mathbf{g} + \mathbf{q}. \tag{6.45}$$

There is a slight difference in the definition of our $G(q)$ and that of Cochran, which might cause confusion when referring to the original article. The relation between the Cochran $G(q)$ and our own is therefore given in the following equation:

$$[G(Q)]_{\text{Cochran}} = -(\Omega_0 q^2/2\pi e^4)G(Q) \tag{6.46}$$

It is interesting to observe that in going from the experimentally determined $E_{ij}(q)$ to a pair interaction via the function $G(q)$ (see Eqs. (6.43) and (6.1)) we have no need of the dielectric constant $\varepsilon(q)$. This eliminates a major source of uncertainty, although we have at the same time retained the linear screening approximation. If on the other hand the final result sought is the bare ion form factor or screened ion potential, some form for $\varepsilon(q)$ has to be assumed.

Cochran used the experimental phonon spectra of Woods et al. [30] for sodium to obtain values of $G(Q)$ and the ion–electron–ion interaction. In practice $G(Q)$ becomes very small beyond the nearest reciprocal lattice point. By considering phonons propagating in each of the three symmetry directions $\langle 100 \rangle$, $\langle 110 \rangle$, and $\langle 111 \rangle$ the frequency Eq. (6.35) is greatly simplified. For example, the $\langle 100 \rangle$ branch of the curve has two independent frequencies, one longitudinal and one transverse:

$$m\omega_{\mathrm{L}}^2(\mathbf{q}) = M_{ii}(\mathbf{q})$$
$$m\omega_{\mathrm{T}}^2(\mathbf{q}) = M_{jj}(\mathbf{q}) \tag{6.47}$$

To obtain E_{ii} and E_{jj} it is necessary to subtract the contributions to M_{ii} arising from the closed-shell repulsion, and the direct Coulomb interaction. The former may be calculated by means of the force constants, assuming a Born–Mayer (BM) repulsion between ionic shells: it is at any rate very small for Na. The Coulomb contribution is a standard result [31], arrived at by an Ewald summation procedure. When Cochran calculated $G(Q)$ from the dispersion curves, he found that the different $G(Q)$ obtained from different symmetry directions fell on one smooth curve to within about 1 %. This curve is reproduced in Fig. 6.12.

The pair interaction potential resulting from this $G(Q)$ contains a minimum at about the nearest-neighbor separation and one positive maximum at about the third-neighbor distance. In Fig. 6.10 it is compared to the MYD potential for Na (see Section 6.7). When we examine the very distinct difference between these two interionic potentials, both of which reproduce extremely well the experimental phonon dispersion curves, it becomes evident that while this may be a necessary condition to be satisfied by a pair interaction, it is not sufficient to define precisely the potential.

An increasing amount of evidence for this statement has appeared in recent work, as we shall see in the next section, where we look at the influence of various factors which go to make up the real space potential. In an investigation of the effect of the choice of bare ion potential on the computed phonon spectra for Al, Wallace [32] chose as two different ion potentials a modified point ion form and a local HAA form, and added a BM repulsion. The potential parameters were adjusted to give the best overall fit to the measured phonon frequencies. Dielectric screening was included in the

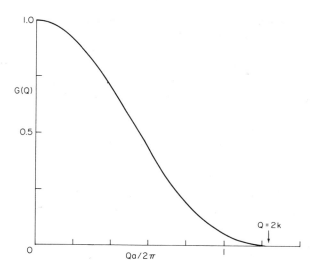

Fig. 6.12. Energy-wavenumber characteristic $G(Q)$ for Na obtained by Cochran from phonon dispersion curves. The Cochran $G(Q)$ is slightly different from that used in this text (see Eq. (6.44) and a is the atomic unit. (After Cochran [29].)

Hubbard–Sham approximation. The BM contribution was found to be negligible. The parameters of the two potentials which gave the best fit to the experimental curves yielded screened ion form factors which were quite different for $q \geq 2k_F$. This large q region however represents an important contribution to the phonon frequencies—it was necessary in Wallace's calculation to take the summation to at least $14k_F$. Thus two essentially different ion potentials reproduce very satisfactorily the experimental phonon spectra.

6.10 Sensitivity of the Potential to the Choice of $w_b(q)$ and $\varepsilon(q)$

It is by now evident that the form of the real-space pair potential, and in particular the magnitude of the long-range oscillations, depend on the model for the screened ion pseudopotential. It is natural therefore to ask to what extent the choice of the bare ion potential w_b and the dielectric screening function $\varepsilon(q)$ separately influence $V(r)$ and to decide whether a choice may be made among the different forms available.

To estimate the sensitivity of $V(r)$ to these two aspects of the model, consider the basic Eqs. (6.1) and (6.2) for the pair interaction. Since the function $G(q)$ contains a factor $[1 - 1/\varepsilon(q)]$ which is very sensitive to the precise form of $\varepsilon(q)$, the choice of the electron–electron contribution to $\varepsilon(q)$ could be quite significant in determining the final potential. The form of $w_b(q)$ determines first the steepness and position of the hard core part of $V(r)$ inside the first minimum. Also, since the long-range oscillations are caused by the logarithmic singularity in $\varepsilon(q)$ at $q = 2k_F$, the amplitude of these oscillations will depend on the magnitude of $w_b(2k_F)$.

The problem of sensitivity of $V(r)$ to the model has been the subject of some recent investigations [33–35] in which the potential $V(r)$ for sodium metal was computed for different forms of w_b and ε. The results may be summarized in the following way:

1. When both local and nonlocal bare ion potentials are screened in the Hartree approximation, the resulting pair potential has no first minimum in the nearest-neighbor position.

2. When many-electron effects are included in $\varepsilon(q)$ the nearest-neighbor minimum is present, but its depth is extremely sensitive to the model chosen, even for dielectric functions which do not differ greatly in form.

3. The amplitude of the oscillations depends sensitively on the point at which the bare ion form factor crosses the q-axis (which determines its value at $q = 2k_F$). This tends to cast doubt upon the more simple empirical models.

4. Widely differing potentials reproduce with good accuracy the experimental phonon dispersion curves, indicating that the latter provide an insufficient test of the potential (see also Section 6.9).

The authors of these investigations conclude that it is important to choose the best available form for both $w_b(q)$ and $\varepsilon(q)$ in order to minimize uncertainty in the pair interaction. The question of which forms are the best is not likely to be the subject of a concensus agreement, besides which the field is in a state of rapid development. At the present time, however, it seems reasonable to suggest that a combination of the Shaw optimized model potential [36] and the treatment of screening by Singwi *et al.* [37] will produce the most reliable interatomic potential presently available. Such a potential is described in a recent paper by Shyu *et al.* [35].

6.11 SOME OBSERVATIONS CONCERNING INTERATOMIC POTENTIALS AND PSEUDOPOTENTIAL THEORY

Before completing this section of the book devoted to pseudopotential theory and its applicability to the interaction between atoms and ions, it would perhaps be useful to note some of the advantages and disadvantages of the method. We list below some very basic points regarding pair interactions derived on the basis of pseudopotential theory:

1. They are valid only under conditions of constant volume, since they are derived from the band-structure energy term in the equation for the total energy of the ion–electron system (see Eq. (5.49)), ignoring the first two volume-dependent terms of this equation. Thus they may be applied only to small variations of ion position keeping the volume constant.

2. They do not apply to situations where the ion cores touch or overlap leading to a closed-shell repulsion. In the small core approximation of pseudopotential theory the core wavefunctions are unaltered in the solid with respect to those of the free ion, which would not be the case should the cores overlap. The pair interaction breaks down at interatomic distances significantly less than the equilibrium separation in the solid, owing to ion core distortion effects. Consequently, pseudopotential pair interactions are of very limited use in situations far removed from equilibrium, such as for example atomic collisions. Then we must rely on some type of repulsive potential such as was described in Chapters II–IV.

3. They are restricted to the so-called normal metals, where the small core approximation holds true. This is the basic difficulty in applying the theory to the noble and transition metals where the large d-shells overlap in the metal at equilibrium. The best results are achieved for the alkalis, where the ion cores are well separated in the crystalline state. Even of the normal metals, those which possess fairly large cores, such as Al, require careful consideration. It is not inconceivable that some success will be achieved in the near future in applying the theory to the noble metals, taking into account the effects of touching d-shells (see Section 5.10). Such an advance would be of great benefit in view of the greater practical application of these metals.

4. Unless their development contains some experimental information usually in the form of adjustable parameters, pseudopotential pair interactions are of limited success in application to real physical situations. We are not yet capable of calculating the potential directly and reliably from first principles even for the simplest metals. Rather we must use the potential as an intermediate theoretical step between one type of experiment and another. If we adjust the potential parameters to one clear and unambiguous set of experimental results we may then go on to use the resulting potential to help clear up an interpretational ambiguity in the conclusions of experiments involving a separate physical property. This does not of course preclude the future theoretical development of reliable potentials after some further refinements in the model.

5. The electron screening contribution to the ion–ion potential becomes more influential the larger the valence, so that any errors involved in the linear screening approximation will be more significant in the polyvalent than in the monovalent metals. In a similar manner the use of a purely local or semilocal pseudopotential is less justified for the polyvalent metals. The

Friedel oscillations in the interatomic potential at large separations are caused by the singularity in the dielectric constant at $q = 2k_F$. It is interesting to note in this respect that where the pseudopotential parameters have been adjusted to experiment in the potential these oscillations become very small. It is reasonable to suggest that at least for the alkali metals the large-r oscillations are unimportant. This is at least partly confirmed by the fact that the presently available phonon dispersion curves for the alkalis show no discernible Kohn anomaly [38].

    6. The form of the real-space potential at distances in the region of and beyond the nearest neighbor separation is rather sensitive to the choice of both the bare ion model and the type of electron gas screening function. Since all the choices presently available involve some form of approximative treatment of the complicated many-body problem, this sensitivity should be borne in mind when applying such a pair interaction to atomic problems.

6.12   CONCLUSION

    Although the pseudopotential theory is far from being fully developed and refined, it has nonetheless cleared the way for a great deal of progress during the past decade in the theory of metals. Reasonably good approximations to the interatomic potentials have become available for the simple metals in the region of and beyond the nearest-neighbor separation in the solid. For many atomic problems such as resistivity, elastic constants etc., it is not necessary to go as far as the final Fourier inversion to real space and it is often more convenient to work entirely in wavenumber space. There are however a number of instances such as defect studies where it is more useful to examine the problem in real space through the pair interatomic potential.

REFERENCES

1. R. Pick, CEA (France) Rep. No. CEA-R-2820 (1965).
2. W. A. Harrison, "Pseudopotentials in the Theory of Metals." Benjamin, New York, 1966.
3. I. V. Abarenkov and V. Heine, *Phil. Mag.* **12**, 529 (1965).
4. M. H. Cohen, *J. Phys. Radium* **23**, 643 (1963).
5. F. S. Ham, *Solid State Phys.* **1**, 127 (1955).
6. A. E. O. Animalu and V. Heine, *Phil. Mag.* **12**, 1249 (1965).
7. W. M. Shyu and G. D. Gaspari, *Phys. Rev.* **163**, 667 (1967).
8. A. O. E. Animalu, F. Bonsignori, and V. Bortolani, *Nuovo Cimento* **44**, 159 (1966).
9. I. M. Torrens and M. Gerl, unpublished work.
10. R. W. Shaw, *J. Phys. C* **2**, 2335 (1969).
11. R. W. Shaw and R. Pynn, *J. Phys. C* **2**, 2071 (1969).
12. A. O. E. Animalu, *Phil. Mag.* **11**, 379 (1965).

13. L. J. Sham and J. M. Ziman, *Solid State Phys.* **15**, 221 (1963).
14. N. W. Ashcroft, *Phys. Lett.* **23**, 48 (1966).
15. W. Geldart and S. H. Vosko, *Canad. J. Phys.* **44**, 2137 (1966).
16. N. W. Ashcroft, *J. Phys. C* **1**, 232 (1968).
17. R. A. Cowley, A. D. B. Woods, and G. Dolling, *Phys. Rev.* **150**, 487 (1966).
18. W. M. Shyu and G. D. Gaspari, *Phys. Rev.* **170**, 687 (1968).
19. P. S. Ho, *Phys. Rev.* **169**, 523 (1968).
20. P. S. Ho, private communication 1969.
21. J. Friedel, *Phil. Mag.* **43**, 153 (1952).
22. A. Meyer, C. W. Nestor, Jr., and W. H. Young, *Proc. Phys. Soc.* **92**, 446 (1967).
23. F. Hermann and S. Skillman, "Atomic Structure Calculations." Prentice-Hall, Englewood Cliffs, New Jersey, 1963.
24. A. Meyer, W. H. Young, and J. M. Dickey, *J. Phys. C* **1**, 486 (1968).
25. J. E. Inglesfield, *J. Phys. C* **2**, 1285, 1293 (1969); *Acta Met.* **17**, 1395 (1969).
26. M. D. Johnson, P. Hutchinson, and N. H. March, *Proc. Roy. Soc.* **A282**, 283 (1964).
27. A. Paskin and A. Rahman, *Phys. Rev. Lett.* **16**, 300 (1966).
28. O. P. Gupta, *Phys. Rev.* **174**, 668 (1968).
29. W. Cochran, *Proc. Roy. Soc.* **A276**, 308 (1963).
30. A. D. B. Woods, B. N. Brockhouse, R. H. March, A. T. Stewart, and R. Bowers, *Phys. Rev.* **128**, 1112 (1962).
31. M. Born and K. Huang, "Dynamical Theory of Crystal Lattices." Oxford Univ. Press, London and New York, 1954.
32. D. C. Wallace, *Phys. Rev.* **187**, 991 (1969).
33. W. M. Shyu and G. D. Gaspari, *Phys. Lett.* **30A**, 53 (1969).
34. M. S. Duesbery and R. Taylor, *Phys. Lett.* **30A**, 496 (1969).
35. W. M. Shyu, J. H. Wehling, M. R. Cordes, and G. D. Gaspari *Phys. Rev.* **B4**, 1802 (1971).
36. R. W. Shaw, *Phys. Rev.* **174**, 769 (1968).
37. K. S. Singwi, M. P. Tosi, R. H. Land, and A. Sjolander, *Phys. Rev.* **176**, 589 (1968); *Phys. Rev. B* **1**, 1044 (1970).
38. W. Kohn, *Phys. Rev. Lett.* **2**, 393 (1959).

CHAPTER VII

ATOMIC COLLISION THEORY
AND INTERATOMIC POTENTIALS

7.1 ATOMIC COLLISIONS

We return now to the higher energy range of atomic interactions, where the atoms are no longer in equilibrium or near-equilibrium in a solid or liquid, but are undergoing dynamic collisions with energies in the region of $\sim 10^2$–10^5 eV. At the higher energy end of this range there are inevitably energy losses due to electronic excitation or ionization. We shall neglect them in this consideration since, as we shall see in Section 7.8, it is quite justifiable to separate, to a first approximation, elastic and inelastic aspects of any two-body collision. Furthermore, all collisions will be considered to obey the laws of classical mechanics, which is a valid assumption for the energy and mass range considered here, except perhaps for glancing collisions with very low energy transfers.

Atomic collision phenomena are basic to many branches of modern physics, and it must be emphasized that in neglecting inelastic energy loss effects such as excitation, ionization, or charge transfer we are voluntarily limiting ourselves to one small corner of a vast field. This corner is nevertheless of vital importance in the treatment of problems of radiation damage in solids and collisions in gases. It also applies to theoretical studies of techniques such as the sputtering of crystal surfaces by charged particles, and the injection of energetic ions into crystals by channeling between lattice lines or planes.

The classical theory of scattering from a central force-field is standard and well documented [1]. Depending on the nature of the force-field, scattering angles and cross sections may be obtained with varying difficulty either analytically or numerically. A number of procedures exist for simplifying the scattering integral in order to solve it analytically. Some of these we shall discuss following a brief summary of the general theory. They consist in the main of approximating or "matching" the central potential for the purpose of assuring convergence and facilitating the problem. The scattering data thereby obtained may be used in wider applications such as the slowing down of charged particles in solids, assuming a sequence of two-body collisions. Then given an interatomic potential a theoretical picture may be constructed for any event involving elastic atomic collisions, which may subsequently be compared with experimental data.

Alternatively, it is quite feasible to reverse the above procedure, using experimental data on some phenomenon containing atomic collisions to determine the interatomic potential. In practice it is necessary to have quite precise information on the scattering processes involved in order to establish the interaction. Information on slowing down of atoms in solids would be unsatisfactory in view of the large number of unknown variables influencing the observable results. The best form of experiment consists of the scattering of a well-defined beam of monoenergetic atoms or ions from an assembly of atoms or ions in a scattering chamber, or indeed from another well-defined beam.

Potential determinations from these scattering data may make use of the normal scattering theory using adjustable parameters in an assumed form of two-body interaction, and adjusting these until they reproduce experimental cross sections. No modifications to the theory are required to follow this procedure. It is however possible to solve the "inversion problem" and derive a theoretical expression for the interatomic potential starting from known scattering cross sections. We shall describe this form of approach in the latter half of this chapter, going on in the following chapter to give an account of experiments which can provide the required scattering data.

Since the concept of an interatomic potential for a two-body collision process requires a classical or semiclassical description of the collision, it is appropriate to investigate the limits of the validity of the classical approximation. It has been shown [2] that there is a critical scattering angle θ_c below which quantum mechanical effects must be taken into consideration. This angle is related to the de Broglie wavelength of the colliding atoms, and may be written

$$\theta_c \simeq \lambda/2R_0 = \pi\hbar/\mu v R_0 \tag{7.1}$$

In this equation R_0 is the distance of closest approach, v the relative velocity before the collision, and μ the reduced mass $(\mu = m_1 m_2/(m_1 + m_2))$. To illustrate this numerically, for two He atoms colliding with relative energy 10 eV, θ_c is about $0.5°$, and for a 1000 eV collision θ_c becomes $\sim 0.05°$. For heavier particles it is easily seen from Eq. (7.1) that θ_c is smaller in inverse proportion to the reduced mass. So a classical treatment seems to be quite in order for energetic atom collisions. For collisions involving thermal energies, on the other hand, θ_c is about $10°$ in the case of a He–He interaction. Thus a quantum approach would be necessary. We shall not be concerned here with such low-energy collisions from the interatomic potential viewpoint, and will therefore omit any quantum effects.

7.2 Reduction of the Two-Body Scattering Problem

Under a purely central potential, the scattering of one particle on another can be reduced to an equivalent problem of the motion of a single particle in a fixed central force-field, which greatly simplifies the mathematical treatment. The proof of this is quite elementary. Suppose we have two particles of masses

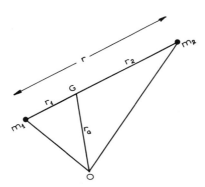

Fig. 7.1. Two-particle system with reference to a fixed origin O and center of mass G.

m_1 and m_2 at positions defined by $\mathbf{r}_1$ and $\mathbf{r}_2$ with respect to their center of mass G (see Fig. 7.1). We assume that their motion is influenced only by their mutual interaction. We can write down the equation for the kinetic energy E_k of the system which moves relative to a fixed origin O:

$$E_k = \tfrac{1}{2}m_1\dot{\mathbf{r}}_1{}^2 + \tfrac{1}{2}m_2\dot{\mathbf{r}}_2{}^2 + \tfrac{1}{2}(m_1 + m_2)\dot{\mathbf{r}}_G{}^2 \tag{7.2}$$

The energy E_k is composed of the motion of the two particles relative to their center-of-mass and the motion of the center of mass relative to the origin. But we can express $\mathbf{r}_1$ and $\mathbf{r}_2$ in terms of the vector $\mathbf{r}$ separating m_1 and m_2:

$$\mathbf{r}_1 = -(\mu/m_1)\mathbf{r}$$
$$\mathbf{r}_2 = (\mu/m_2)\mathbf{r} \tag{7.3}$$

Substituting Eq. (7.3) into (7.2) gives

$$E_k = \tfrac{1}{2}\mu\dot{\mathbf{r}}^2 + \tfrac{1}{2}(m_1 + m_2)\dot{\mathbf{r}}_G{}^2 \tag{7.4}$$

But since there is no force acting on the two-particle system considered as a whole, the center-of-mass must be either at rest or moving along a straight line, so $\dot{\mathbf{r}}_G$ will be constant. The first term of Eq. (7.4) is equivalent to that for a single particle of mass μ moving about a fixed center of force with vector $\mathbf{r}$, in a plane if the force is central.

The first stage in the two-body scattering problem is thus to reduce the absolute motion in space to the relative motion of the particles in the center-of-mass (CM) system. Figure 7.2 illustrates the practical example of a collision between a moving particle of mass m_P and an initially stationary particle m_T, first in the laboratory (L) system and then in the CM system. The

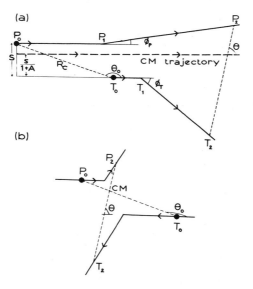

Fig. 7.2. Classical collision of a projectile P with a target T, shown in (a) the laboratory or L system, and (b) the center-of-mass or CM system. Only the asymptotes of the trajectories are shown in the figure. The notation is explained in the text.

diagram shows the *impact parameter s*, which is the length of the perpendicular between the initial position T_0 of the target and the incident trajectory of the projectile $P_0 P_1$. This as we shall see is an important quantity in the scattering process, defining the hardness of the collision. The motion of the center of mass for this collision is also illustrated in the figure.

If the potential between the two particles is $V(r)$, then following Eqs. (7.2)–(7.4) the collision may be thought of in terms of a single mass $\mu = m_P m_T/(m_P + m_T)$, moving under the central force $-dV/dr$ acting along the separation vector $\mathbf{r}$. In order to obtain the scattering angles in the L system we must first determine the equation of the orbit of μ in the CM system. Representing its motion vector $\dot{\mathbf{r}}$ by polar coordinates (r, θ), we can write down its angular momentum l, which must remain constant since the force depends only on the magnitude of $\mathbf{r}$.

$$l = mr^2 \dot{\theta} \tag{7.5}$$

Similarly the total energy of the system E, again a constant of the motion, is

$$E = \tfrac{1}{2}\mu(\dot{r}^2 + r^2\dot{\theta}^2) + V(r) \tag{7.6}$$

where the first term is the kinetic energy.

Substituting for $\dot{\theta}$ in Eq. (7.6) using (7.5), we obtain

$$\dot{r} = \{(2/\mu)[E - V(r) - l^2/2\mu r^2]\}^{1/2} \tag{7.7}$$

Since, from (7.5) $d\theta = l/\mu r^2 \, dt$, we may write down the differential equation of the orbit as

$$d\theta = \frac{l \, dr}{\mu r^2 \{(2/\mu)[E - V(r) - l^2/2\mu r^2]\}^{1/2}} \tag{7.8}$$

or in integral form, if $\theta = \theta_0$ initially, corresponding to $r = r_0$,

$$\theta - \theta_0 = \int_{r_0}^{r} \frac{l \, dr}{r^2 \{2\mu[E - V(r)] - l^2/r^2\}^{1/2}} \tag{7.9}$$

The time integral of the scattering may be written down directly from Eq. (7.7). If we assume that $t = 0$ initially, then

$$t = \int_{r_0}^{r} \frac{\mu \, dr}{\{2\mu[E - V(r)] - l^2/r^2\}^{1/2}} \tag{7.10}$$

At this point we introduce the impact parameter s, which is related to the initial (and constant) angular velocity:

$$l = s(2\mu E)^{1/2} \tag{7.11}$$

Substituting this expression into the Eqs. (7.9) and (7.10) we have the basic equations normally used in scattering problems:

$$\theta - \theta_0 = \int_{r0}^{r} s\, dr/r^2 f(r), \qquad t = (\mu/2E)^{1/2} \int_{r0}^{r} dr/f(r) \qquad (7.12)$$

where

$$f(r) = \{1 - s^2/r^2 - V(r)/E\}^{1/2} \qquad (7.13)$$

For an unbounded orbit ($\theta_0 = \pi$) the limits of integration are ∞ and the distance of closest approach R_0 of the composite particle to the center of force, defined by the equation

$$f(R_0) = 0 \qquad (7.14)$$

Then the value of θ after the collision is given by

$$\theta_F = \pi - 2 \int_{R_0}^{\infty} s\, dr/r^2 f(r) \qquad (7.15)$$

We note that this equation has a singularity at $r = R_0$, which must be circumnavigated in the integration process. One method is to make the substitution:

$$r = R_0/(1 + u^2) \qquad (7.16)$$

Then

$$\theta_F = \pi - 4s \int_{0}^{1} du/\{s^2(2 - u^2) + (R_0^2/Eu^2)[V(R_0) - V(r)]\}^{1/2} \qquad (7.17)$$

By substituting for u^2 in the second term of the denominator and letting u tend to zero, it is easily verified that the singularity no longer exists when

$$V'(R_0) < 2Es^2/R_0^3 \qquad (7.18)$$

The scattering integral may be evaluated in analytical form for only a very few power law potentials, but a number of approximations which we shall discuss later have permitted an extension of the scope of the theory by semianalytical procedures. For accurate solutions the integration may be easily performed by computer.

We note that in solving a scattering problem such as that represented in Fig. 7.2(a) by reduction to CM coordinates, the energy E of Eq. (7.18) is the relative energy of the two particles, given by

$$E = AE_P/(1 + A) \qquad (7.19)$$

where A is the mass ratio ($= m_T/m_P$). The problem is thus solved in the CM system, and it remains simply to express the scattering angles of projectile and target in laboratory coordinates.

7.3   Transformation to the Laboratory System

The scattering angles ϕ_P and ϕ_T in the L system are easily obtained by referring to the velocity vector diagram after the collision. From Fig. 7.3,

$$\tan \phi_P = \frac{\mathbf{v}_{PCM} \sin(\theta_0 - \theta)}{\mathbf{v}_{PCM} \cos(\theta_0 - \theta) + \mathbf{v}_G} \tag{7.20}$$

But we can express both $\mathbf{v}_{PCM}$ and $\mathbf{v}_G$ in terms of the initial projectile velocity $\mathbf{v}_{P0}$. By conservation of momentum:

$$m_P \mathbf{v}_{PCM} = \mu \mathbf{v}_{P0} \tag{7.21}$$

and

$$(m_P + m_T)\mathbf{v}_G = m_P \mathbf{v}_{P0} \tag{7.22}$$

Then Eq. (7.20) becomes

$$\phi_P = \tan^{-1}\left\{ \frac{A \sin(\theta_0 - \theta)}{1 + A \cos(\theta_0 - \theta_0)} \right\} \tag{7.23}$$

The equivalent equation for the target atom is obtained by substituting m_T for m_P in Eq. (7.21), so that

$$\phi_T = \tan^{-1}\left\{ \frac{\sin(\theta_0 - \theta)}{1 + \cos(\theta_0 - \theta)} \right\} = \frac{\pi}{2} - \frac{\theta_0 - \theta}{2} \tag{7.24}$$

By using a little algebra on the above conservation equations the final energies E_P' and E_T' of the projectile and target may be expressed in terms of

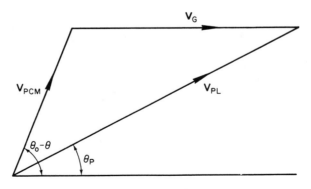

Fig. 7.3. Velocity vector diagram containing CM and L system projectile velocities $\mathbf{v}_{PCM}$ and $\mathbf{v}_{PL}$ as well as the center-of-mass velocity $\mathbf{v}_G$.

the initial projectile energy and the scattering angle as follows:

$$E_P' = \left\{1 - \frac{4A}{(1 + A)^2} \sin^2\left(\frac{\theta_0 - \theta}{2}\right)\right\}E_P$$

$$E_T' = \frac{4A}{(1 + A)^2} \sin^2\left(\frac{\theta_0 - \theta}{2}\right)E_P \qquad (7.25)$$

The energy transfer T $(= E_T')$ is often used instead of the scattering angle as the operating variable. The maximum value of energy transfer T_m occurs when $\sin^2[(\theta_0 - \theta)/2] = 1$, and is simply

$$T_m = [4A/(1 + A)^2]E_P \qquad (7.26)$$

So the relation between scattering angle and energy transfer becomes

$$\sin^2((\theta_0 - \theta)/2) = T/T_m \qquad (7.27)$$

As we shall see later, impact parameters and scattering cross sections are frequently expressed in terms of the energy transfer.

The detailed trajectories of the projectile and target may be found with the aid of the scattering integral and time integral, by assuming a collision cutoff R_c beyond which the interaction of the two particles is assumed negligible. The only difference is that now the initial conditions for these integrals have changed

$$\theta_0 = \pi - \tan^{-1}\left\{\frac{s}{(R_c^2 - s^2)^{1/2}}\right\}, \qquad r_0 = R_c \qquad (7.28)$$

The evaluation of the time integral t of Eq. (7.12) enables the positions of the projectile, target and CM to be localized at any time during the collision, in particular at the end of the interaction when r is once again equal to R_c.

An alternative description of the course of the event employs the incoming and outgoing asymptotes of the trajectories, which are represented on Fig. 7.2(a). The former are trivial ($y = s$ and $y = 0$). The outgoing asymptote is defined as the line tangential to the trajectory at the point where the collision ceases ($r = R_c$), and may be found for the projectile and target by simple geometry. It may be verified that the outgoing asymptote of the target passes through the intersection of the two asymptotes of the projectile. The particular convenience of the asymptote representation lies in its application to the computation of a number of sequential two-body collisions. Instead of placing the projectile at P_2 and the target at T_2 after the collision, they are placed at P_1 and T_1 respectively, with velocity vectors along P_1P_2 and T_1T_2. This procedure eliminates the risk of neglecting a second collision which might take place before the projectile reaches P_2 or the target (now a second projectile) T_2.

It is well-nigh impossible to measure experimentally the results of a single two-body atomic collision. We must instead approach the problem statistically, reducing to a minimum the number of variables. We cannot localize the target atom to any greater extent than is defined by the beam area and the macroscopic volume in which the collision takes place. The incident particle energy, mass, and direction of motion may be well defined and the post-collision particle detectors may be very selective in all these quantities. Then to analyse the experimental results it is necessary to introduce the concept of probabilities of scattering with a given energy transfer and angular deflection. These *scattering cross sections*, as they are termed, rely on a large number of two-body collisions to provide the necessary statistics. We shall go on to consider these cross sections in Section 7.5, but first we discuss briefly an approximation to the scattering integral which may be used when a collision is not too violent.

7.4   The Impulse Approximations

If the collision between two particles is sufficiently " soft," and the angle of deflection of the incoming particle small, calculation of the full scattering integral may be side-stepped by approximative methods. The simplest of these is the *impulse approximation* or *momentum approximation*, which in the zero-order assumes no deflection of the incident particle. The energy transferred to the target is calculated by assuming that it receives an impulse integrated over the whole straight-line trajectory of the projectile. The situation is illustrated by Fig. 7.4. It may be easily calculated that the energy

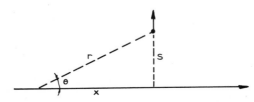

Fig. 7.4. Impulse approximation collision. The target moves off perpendicular to the incident direction of the projectile.

transfer T resulting from this impulse is given in terms of the interatomic potential and the impact parameter by

$$T = (1/E_P) \int_s^\infty (s/(r^2 - s^2)^{1/2})(-dV/dr)\, dr \qquad (7.29)$$

If we assume a BM potential for example, of the form

$$V(r) = A\, e^{-r/a} \qquad (7.30)$$

then the energy transfer becomes

$$T = (A^2 s^2/E_P a^2)[K_0(s/a)]^2 \qquad (7.31)$$

Taking the target momentum as being in a direction perpendicular to the incident projectile trajectory, we may calculate an angular deflection for the projectile

$$\phi_P = (As/E_P a)K_0(s/a) \tag{7.32}$$

This first-order approximation is evidently only applicable to extremely glancing collisions. Lehmann and Leibfried [3] have extended the theory to higher orders by expanding the classical scattering ntegral in powers of the potential, in order to calculate analytically the angular deflection $(\theta - \theta_0)$. They performed the integration using a complex-plane contour method for two potentials, namely the BM form of Eq. (7.30) and an exponentially screened Coulomb form:

$$V(r)^C = A(a/r)e^{-r/a} \tag{7.33}$$

We shall not describe the detailed mathematical analysis, but merely quote the results of Lehmann and Leibfried. The different orders of approximation are the result of cutting off the power series for the potential at different stages. Taken as far as third order the angular deflections ϕ^{BM} and ϕ^C for the two potentials (7.30) and (7.33) are as follows

$$\phi_1^{BM} = \frac{A}{E}\frac{s}{a} K_0\left(\frac{s}{a}\right) \tag{7.34}$$

$$\phi_2^{BM} = -\frac{A^2}{E^2}\left\{\frac{s^2}{a^2} K_1\left(\frac{2s}{a}\right) - \frac{3s}{2a} K_0\left(\frac{2s}{a}\right)\right\} \tag{7.35}$$

$$\phi_3^{BM} = \frac{9}{8}\frac{A^3}{E^3}\left\{\frac{s^3}{a^3} K_0\left(\frac{3s}{a}\right) - \frac{8s^2}{3a^2} K_1\left(\frac{3s}{a}\right) + \frac{5s}{3a} K_0\left(\frac{3s}{a}\right)\right\} \tag{7.36}$$

$$\phi_1{}^C = \frac{A}{E} K_1\left(\frac{s}{a}\right) \tag{7.37}$$

$$\phi_2{}^C = -\frac{A^2}{E^2} K_1\left(\frac{2s}{a}\right) \tag{7.38}$$

$$\phi_3{}^C = \frac{9}{8}\frac{A^3}{E^3}\left\{K_1\left(\frac{3s}{a}\right) - \frac{a}{3s} K_0\left(\frac{3s}{a}\right) - \frac{2a^2}{9s^2} K_1\left(\frac{3s}{a}\right)\right\} \tag{7.39}$$

In a numerical investigation of scattering angles predicted by the impulse approximation for various energies and impact parameters, Lehmann and Leibfried were able to describe a range of validity of the impulse approxima- tion for these two potentials. This permitted region is best illustrated by reference to Fig. 7.5, in which the curves of the original paper are reproduced. These represent a good guide in that they are generally applicable to any

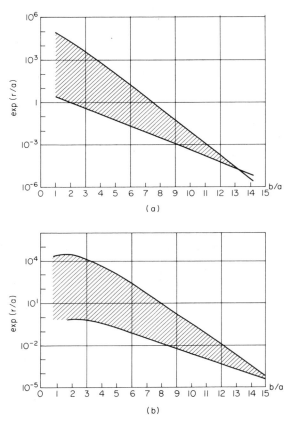

Fig. 7.5. The shaded area represents the region of validity of the momentum approximation, for (a) the BM potential, and (b) the Bohr potential (see Eqs. (7.30) and (7.33)). In this figure the impact parameter is denoted by b. (After Lehmann and Leibfried [3].)

collision between any two particles under a potential similar in form to one of the two chosen. The advantage of using the impulse approximation is of course that it avoids a numerical integration of the scattering equation. As a further numerical indication of its range of application we compare in Tables 7.1a and 7.1b the scattering angle as calculated from the third-order impulse approximation with that obtained by numerical integration of the full scattering integral. The colliding particles are Cu atoms of two relative energies (5 and 10 keV) interacting with the GGMV potential 2 (see Section 4.9). The impact parameter is given in units of the lattice parameter of Cu. We see that less than 1% divergence between the two results is evident at impact parameters greater than 0.3 for energy 5 keV and 0.2 for 10 keV.

TABLE 7.1

COMPARISON OF SCATTERING ANGLES FOR Cu–Cu COLLISIONS PREDICTED BY THE
FULL SCATTERING INTEGRAL AND BY THE IMPULSE APPROXIMATION

TABLE 7.1[a]

| s (units of a_L) | CM angular deflection (degrees) | | Difference |
	Impulse approx.	Scattering int.	
0.001	528.804	177.924	350.880
0.01	1180.77	159.464	1021.31
0.05	109.526	95.096 1	14.430 3
0.1	40.385 9	52.928 2	−12.542 3
0.2	17.100 4	16.857 5	0.242 874
0.3	4.592 69	4.588 79	0.390 278D − 02
0.4	1.082 24	1.082 16	0.733 498D − 04
0.5	0.237 042	0.237 042	0.717 169D − 06
0.6	0.501 777D − 01	0.501 791D − 01	−0.142 544D − 05
0.7	0.104 314D − 01	0.104 316D − 01	−0.157 661D − 06

[a] Relative energy of colliding atoms = 5000.0 eV.

TABLE 7.1b[a]

| s (units of a_L) | CM angular deflection (degrees) | | Difference |
	Impulse approx.	Scattering int.	
0.001	84.309 6	175.257	−90.947 3
0.01	213.844	136.050	77.794 4
0.05	66.594 3	60.957 1	5.637 19
0.1	30.900 2	31.865 9	−0.965 690
0.2	9.346 56	9.330 84	0.157 244D − 01
0.3	2.385 20	2.385 00	0.198 547D − 03
0.4	0.547 291	0.547 264	0.271 230D − 04
0.5	0.118 866	0.118 865	0.332 058D − 06
0.6	0.251 062D − 01	0.251 069D − 01	−0.713 024D − 06
0.7	0.521 652D − 02	0.521 660D − 02	−0.788 571D − 07

[b] Relative energy of colliding atoms = 10000.0 eV.

7.5   SCATTERING CROSS SECTIONS

The differential scattering cross section for a given direction based on an incident beam of particles of known intensity I_0 is defined as the fraction of the incident intensity scattered into unit solid angle about this direction. It is normally expressed in terms of the planar scattering angle rather than the solid angle Ω:

$$I(\Omega) \, d\Omega / I_0 = \sigma(\Omega) \, d\Omega = 2\pi\sigma(\phi) \sin(\phi) \, d\phi \qquad (7.40)$$

Here we assume that the density of scattering centers has azimuthal symmetry. Since ϕ depends only on the incident beam energy E_0 and the impact parameter s, then the number of particles scattered through angle ϕ into the solid angle $d\Omega$ is just the number which have impact parameter between s and $s + ds$. Thus

$$2\pi\sigma(\phi) \sin \phi \, d\phi = -2\pi s \, ds \qquad (7.41)$$

or

$$\sigma(\phi) \sin \phi = -\tfrac{1}{2} \, ds^2/d\phi \qquad (7.42)$$

The impact parameter provides the link between the interatomic potential and the scattering cross section. This is true both for the calculation of $\sigma(\phi)$, assuming a given potential form, and in the inversion problem where experimental cross sections permit a determination of the two-body interaction.

Integration of Eq. (7.42) provides the relation between the impact parameter and the differential scattering cross section:

$$s^2(\phi) = 2 \int_{\phi}^{\pi} \sigma(\phi) \sin \phi \, d\phi \qquad (7.43)$$

Linked with the other relation between s and the CM scattering angle θ of Eq. (7.15), we have an effective means of passing between $V(r)$ and $\sigma(\phi)$.

We note that instead of the angular deflection ϕ we might choose the energy transfer T as an equivalent variable. Then Eq. (7.43) becomes

$$s^2 = (1/\pi) \int_{T}^{T_m} \sigma(T) \, dT \qquad (7.44)$$

or

$$\sigma(T) = -\pi \, ds^2/dT \qquad (7.45)$$

where as before T_m is the maximum possible energy transfer, occurring for back-scattered particles.

The *total scattering cross section* S, being the total number of particles scattered into all solid angles per unit incident intensity, is found by integrating Eq. (7.40) over all possible values of ϕ:

$$S = 2\pi \int_{0}^{\pi} \sigma(\phi) \sin \phi \, d\phi \qquad (7.46)$$

This turns out to be divergent for differential cross sections evaluated from scattering potentials of infinite range, and can thus be determined only for potentials artificially cut off at some finite value of r. Comparing Eqs. (7.43) and (7.46):

$$S = \pi s^2(0) \tag{7.47}$$

But if $V(r)$ extends to infinity all incident particles out to $s = \infty$ are deflected by a finite amount. Hence $s^2(0)$ and consequently $S(0)$ are infinite. It is important however to note that changing to a quantum mechanical treatment of small angle collisions will lead to a finite total cross section.

7.6 LARGE-ANGLE SCATTERING USING MATCHING POTENTIALS

Although the classical scattering integrals must in general be solved numerically for any physically realistic interaction potential, there is a limited number of simplified potentials for which it is possible to evaluate the integrals in closed form. For these potentials analytical expressions for scattering angles and scattering cross sections may be derived. For large angle or low impact parameter collisions between atoms most of the effect is felt near to the distance of closest approach. It might therefore be imagined that the scattering could be adequately described by one of these simplified potentials " matched " in value and slope to the true potential at this distance. In this section we shall describe in outline only some of the work which has been done in the field of matching potentials. The technique is primarily of use in scattering theory and analytical treatment of collision processes in solids. It is only tangential to the subject of interatomic potentials, requiring an *a priori* knowledge of the interaction.

Some simple forms of potential were considered by Leibfried and Oen [4]. For high energy collisions occurring at steep parts of the $V(r)$ curve, the hard-sphere approximation might be matched at the distance of closest approach, which depends on the relative energy of the colliding particles:

$$V_m[r < R_0(E)] = \infty$$
$$V_m[r > R_0(E)] = 0 \tag{7.48}$$

Expressed in terms of the energy transfer, this gives very simple expressions for the impact parameter and scattering cross section $\sigma(T)$:

$$s^2(T) = R_0^2(1 - T/T_m) \tag{7.49}$$

$$\sigma(T) = \pi R_0^2/T_m \tag{7.50}$$

Since in fact the slope has not been matched at R_0 for the hard-sphere approximation, this is only valid at very small R_0 where dV/dr is large. An improved form suggested by Leibfried and Oen was a truncated Coulomb matching potential:

$$V_m(r < a') = A(a'/r - 1)$$
$$V_m(r > a') = 0$$
(7.51)

The constants A and a' were obtained by matching $V_m(r)$ and $V_m'(r)$ to the real potential at the distance of closest approach $R_0(E)$. For this potential the impact parameter and scattering cross section are

$$s^2(T) = \left(\frac{Aa'}{2E}\right)^2 \cdot \frac{1 - T/T_m}{(1 + A/E)(T/T_m) + (A/2E)^2}$$
(7.52)

$$\sigma(T) = \frac{\pi}{T_m} \left\{ \frac{Aa'}{2E} \cdot \frac{1 + A/2E}{(1 + A/E)(T/T_m) + (A/2E^2)} \right.$$
(7.53)

We note that it is not essential to match the potentials at the distance of closest approach in a *head-on* collision. The value of $R_0(s)$ for any given s may be found simply by solving the Eq. (7.14). Then the potential may be matched at $R_0(s)$ instead of $R_0(0)$. This however offers no particular advantage when dealing with situations such as the slowing down of energetic atoms in solids, where the greatest effect comes from hard large-angle collisions.

The work of Leibfried and Oen was subsequently extended by Lehmann and Robinson [5] to the matching of a family of truncated power potentials of the form:

$$V_n(r < \beta_n) = (\alpha_n/n)[(\beta_n/r)^n - 1]$$
(7.54)

Figure 7.6 illustrates the matched potentials compared to the real potential $V(r)$ for several values of n. Equations (7.48) and (7.51) are special cases of this potential, with $n = -\infty$ and $n = +1$ respectively. The constants α_n and β_n, found by matching the potential and slope at $r = R_0$, are given by the equations

$$\alpha_n = -[nV(R_0) + R_0V'(R_0)]$$
$$\beta_n = R_0[1 + (n/R_0)(V(R_0)/V'(R_0))]^{-1/n}$$
(7.55)

Since we are dealing with purely repulsive potentials, such that:

$$V(r) \geq 0, \qquad V'(r) \leq 0, \qquad 0 < r < \infty$$
(7.56)

the matching procedure is only valid for n obeying the condition

$$nV(r) < -rV'(r)$$
(7.57)

Lehmann and Robinson compared the results for the fractional transferred energy T/T_m at various impact parameters based on a "true" potential $V(r)$ represented by the Bohr screened Coulomb form. Matching was carried

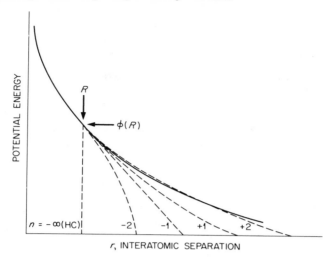

Fig. 7.6. Matching potentials $V_n(r)$ given by Eq. (7.54) for different values of n between $-\infty$ (hard-core) and $+2$, matched to a given "real" potential $\phi(r)$ at $r = R$: (——) $\phi(r)$; (–––)$V_n(r)$. (After Lehmann and Robinson [5].)

out both at $R_0(0)$ and $R_0(s)$ for $n = -\infty$, $n = 1$ and $n = 2$ of the family given by the Eq. (7.54). There was excellent agreement with the results of numerical integration using the Bohr potential when the matching was performed at the same impact parameter s. Of those forms matched at the head-on value of R_0, both the truncated Coulomb ($n = 1$) and the truncated inverse square ($n = 2$) approximations are good over a broad range of s for close collisions, the latter being somewhat superior. For more glancing collisions the energy transfer can be calculated using the momentum approximation [3], which is the best method for large s or small angle scattering.

Following this work, Sigmund and Vajda [6], in a series of reports calculated scattering cross sections for potentials matched to the TF and BM forms of interaction. The matching potentials were truncated Coulomb and a second-order form made up of a sum of the Coulomb and inverse square laws. This could be fitted over a larger range than the simple Coulomb type. Sigmund and Vajda discussed the application of their scattering cross section results to radiation damage cascade theory.

Instead of approximating the interatomic potential and subsequently calculating the exact differential scattering cross section corresponding to the matching potential, it is possible to develop an approximate expression for the differential cross section from the starting point of the exact potential. Several workers [7, 8] have attacked the problem in this fashion, finding differential cross section expressions which may be used in further analytical treatments of multiple collision problems.

There are two principal difficulties inherent in applying matched potentials and their resulting scattering laws to displacement cascade theory. One is their obvious misrepresentation of low-energy transfer collisions, which form a significant part of a cascade. Second, it has been demonstrated by Robinson [9] that the number of displacements in a solid resulting from a primary knocked-on atom is more sensitive than had been previously thought to the form of the scattering laws. It would in fact be equally valid to start from some assumed form of scattering cross section (in Robinson's case depending on T^{-n}) instead of from the interatomic potential itself.

7.7   THE INVERSION PROBLEM

In atomic collision phenomena involving energies greater than a few electron volts we are for all practical purposes justified in assuming a monotonically decreasing interaction potential. In reality, as we have seen in previous chapters, the potential goes through an attractive minimum before increasing towards zero or oscillating at large r. This causes anomalies in low energy collisions, illustrated by the phenomenon of *rainbow scattering* (see Section 8.2). When dealing with collisions in the energy range greater than the displacement threshold in a crystal (~ 25 eV) there is no great inaccuracy introduced by neglecting the attractive part in favor of a form of closed shell (or at high energies screened Coulomb) repulsion. Then there is a unique relationship between the potential, the impact parameter, scattering angle and differential scattering cross section. That is to say, if we assume some form of interatomic potential we may calculate either analytically or numerically the deflection angle or energy transfer for a given impact parameter. This may be compared with experiment or used in another theoretical treatment. Such a procedure has been described in the preceding sections.

Because of the uniqueness of the relation between these scattering variables and the potential, it is conceptually possible to use scattering data at a particular angle or energy transfer to define unambiguously a point on the potential energy curve. One method, as we have mentioned above, is to have some adjustable parameters in an assumed potential form and use the theory in the normal way. While very useful this method is not completely satisfactory, relying as it does on initial assumptions regarding the form of the interaction. The question is, how feasible is it theoretically to invert the problem so as to express the potential in terms of one of the experimentally definable scattering variables and thus evaluate it without recourse to any *a priori* hypothesis? The problem was first resolved by Firsov [10] in an evaluation of some earlier work by Hoyt [11]. Since this is a very important piece of work from the point of view of experimental determination of interatomic potentials, we shall now describe it in some detail.

The starting-point of the theory is the fundamental scattering equation for an unbounded orbit:

$$\theta = \pi - \int_{R_0}^{\infty} \frac{2s}{r\{(1 - V(r)/E)r^2 - s^2\}^{1/2}} \, dr \qquad (7.58)$$

Here θ is the angle of scattering in the CM system. Suppose

$$\psi^2 = (1 - V(r)/E)r^2 \qquad (7.59)$$

Noting that at the distance of closest approach R_0, in accordance with Eq. (7.14) we have simply $\psi = s$, we perform the change of variables in Eq. (7.58):

$$\theta = \int_{s}^{\infty} \frac{d}{d\psi} \left[\ln \frac{\psi^2}{r^2}\right] \frac{s}{(\psi^2 - s^2)^{1/2}} \, d\psi \qquad (7.60)$$

By multiplying both sides of this equation by $ds/(s^2 - \psi^2)^{1/2}$ and integrating from $s = \psi$ to ∞, we obtain with the aid of Dirichlet's theorem:

$$\int_{\psi}^{\infty} \frac{\theta(s) \, ds}{(s^2 - \psi^2)^{1/2}} = \frac{\pi}{2} \ln\left(\frac{r^2}{\psi^2}\right) \qquad (7.61)$$

from which

$$r(\psi) = \psi \exp\left\{\frac{1}{\pi} \int_{\psi}^{\infty} \frac{\theta(s) \, ds}{(s^2 - \psi^2)^{1/2}}\right\} \qquad (7.62)$$

The impact parameter s and the dependence of θ on s are found from the differential scattering cross section $\sigma(\theta)$. Thus, from Eq. (7.43)

$$s^2(\theta) = 2 \int_{0}^{\pi} \sigma(\theta) \sin \theta \, d\theta \qquad (7.63)$$

Equations (7.59), (7.62), and (7.63), taken in reverse order, are the essential steps in the Firsov procedure for determining $V(r)$. As basic data let us assume that we have a curve of the experimental differential scattering cross section $\sigma(\phi)$ as a function of the laboratory scattering angle ϕ for a given constant projectile energy E. The first step is to convert to the CM system so as to obtain $\sigma(\theta)$ as a function of θ and integrate for various θ according to Eq. (7.63). This leads to a table of values of impact parameter s for different scattering angles which by the same token may be interpreted as a table of $\theta(s)$ for various impact parameters s. (It is at this point that one would compare the $\theta(s)$ curve with that calculated from an assumed form of $V(r)$ if one were using the adjustable parameter approach.)

The values of $\theta(s)$ may in principle be used in Eq. (7.62) to obtain $r(\psi)$. A problem however arises because of the singularities in the integral. This may be eliminated by a further change of variable introduced in an application of the Firsov theory to experimental scattering data by Lane and Everhart [12]:

$$u = b/s, \qquad w = b/\psi, \qquad b = Z_1 Z_2 e^2/E \qquad (7.64)$$

Equation (7.62) then becomes

$$r(w) = (b/w) \exp\{(1/\pi) I(w)\} \qquad (7.65)$$

where

$$I(w) = \int_0^w \frac{\theta(u)}{u} \frac{du}{(1 - u^2/w^2)^{1/2}} \qquad (7.66)$$

The last equation may be rewritten

$$I(w) = \int_0^w \frac{\theta(w)(u/w)\,du}{u(1 - u^2/w^2)^{1/2}} - \int_0^w \frac{[\theta(w)(u/w) - \theta(u)]}{u(1 - u^2(w^2)^{1/2}}$$

$$= \theta(w)(\pi/2) - I_1(w) \qquad (7.67)$$

The singularities are thereby removed, since as u approaches zero $\theta(u)$ is less than $\theta(w)(u/w)$ which also approaches zero. Then $I_1(w)$ can be neglected.

The procedure used by Lane and Everhart was to plot θ against u following the $\theta(s)$ data and Eq. (7.64), and to extrapolate the curve to $u = 0$. This last step was necessary owing to lack of data for laboratory scattering angles ϕ between 0 and 1°. This curve is shown schematically in Fig. 7.7. Selecting a

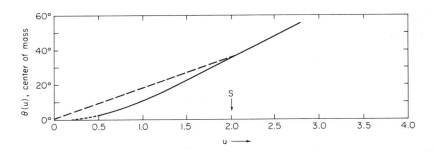

Fig. 7.7. The scattering angle θ is plotted against u (see text for significance of u). The dashed straight line is drawn to $\theta(w)$, where w is a particular value of u. (After Lane and Everhart [12].)

value for w, a straight line $\theta(u) = \theta(w)(u/w)$ was drawn from the origin to the point on the curve corresponding to $u = w$. The difference between the straight line and the curve provided the numerator of the $I_1(w)$ integrand. The integration to obtain $I(w)$ was performed in this manner for a number of values of w, following which it was simple using Eq. (7.65) to find $r(w)$.

Adaptation of Eq. (7.59) to the new variables gives

$$V(r) = E\{1 - b^2/w^2 \cdot 1/r^2(w)\} \tag{7.68}$$

It is then possible to find the potential energy of interaction $V(r)$ for any particular incident particle energy. Note that this is the relative energy in the CM system.

Lane and Everhart compared the potentials obtained from 100-, 50-, and 25-keV ion–atom scattering experiments for inert gases with Firsov and Bohr screened Coulomb potentials (see Section 4.7). The experimental $V(r)$ did not lie on a continuous curve, but in general there was reasonable agreement with the theoretical potentials, although it was not really possible to separate Bohr and Firsov forms. The curves for Ar^+–Ar collisions are reproduced in Fig. 7.8. They noted that other factors than the inadequacy of the theoretical potentials might have intervened to cause the discrepancies. The principal factor was probably experimental error, although inelastic energy losses could possibly cause the discontinuities in the potential curves (see Section 7.8).

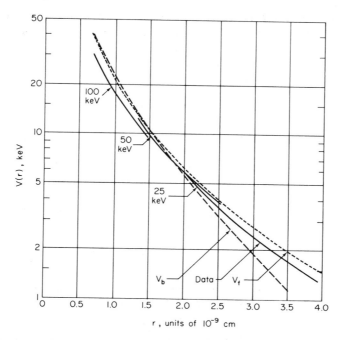

Fig. 7.8. Interatomic potential obtained by Firsov inversion for Ar^+–Ar collisions. There are three nonintersecting solid lines for 100-,50-, and 25-keV incident energies. Also shown are the Bohr (V_b) and Firsov (V_f) screened Coulomb potentials. (After Lane and Everhart [12].)

One of the inconvenient aspects of the Firsov theory is the large amount of experimental work which goes into the determination of the potential curve over a relatively small range of interatomic separation. A complete differential scattering cross section curve is required at the relevant energy E and even then a certain amount of extrapolation is inevitable. Some recent progress in this topic, in the form of the *impact expansion method* [13] permits the use of scattering data at different energies to find a general interatomic potential. The name given to the method stems from similarities between it and the impulse or momentum approximation (see Section 7.4). Without going into the rather complicated mathematical detail of the method, the principle is as follows. The basic variables are the impact parameter s and the *reduced scattering angle* γ, given by

$$\gamma = E\phi(s, E) = \gamma_0(s) + \frac{\gamma_1(s)}{E} + \frac{\gamma_2(s)}{E^2} + \cdots \qquad (7.69)$$

and the *reduced CM scattering cross section* ρ, written

$$\rho = \phi\sigma(E, \phi) \sin \phi \qquad (7.69a)$$

The relation between these new variables (to which must now be added the energy) is sought with the help of Lagrange expansions of the form

$$s(E, \gamma) = \sum_{n=0}^{\infty} \varepsilon^n s_n(\gamma)$$

$$\rho(E, \gamma) = \sum_{n=0}^{\infty} \varepsilon^n \rho_n(\gamma) \qquad (7.70)$$

The Firsov procedure is in fact equivalent to the zero-order terms of these series, where there is no energy dependence. From Eq. (7.43),

$$\frac{-\frac{1}{2} ds_0^2}{d\gamma} = \frac{\rho_0(\gamma)}{\gamma} \qquad (7.71)$$

The zero order inversion, which may be performed as long as $s(\gamma)$ is single-valued, expresses very simply the basic step of the Firsov inversion:

$$\gamma_0(s) = \gamma_0[s_0(\gamma)] \qquad (7.72)$$

Calculations using several potential functions show that the higher terms $\rho_1(\gamma)$, $\rho_2(\gamma)$, etc. almost always decrease rapidly at small γ, whereas $\rho_0(\gamma)$ do not. Thus the set of curves $\rho(E, \gamma)$ for different values of E coincide with $\rho_0(\gamma)$ at small γ. This provided a method of defining $\rho_0(\gamma)$ from data at different energies. Then from Eq. (7.71) $s_0^2(\gamma)$ could be obtained and inverted according to Eq. (7.72) to give $\gamma_0(s)$.

It was found using a Lagrangian form of expansion that $\gamma_n(s)$ could be written in the form

$$\gamma_n(s) = -s \int_{s^2}^{\infty} F_n(x) \, dx/(x - s^2)^{1/2} \qquad (7.73)$$

where $F_n(x)$ is related to the potential $V(r)$ through the equation

$$F_n(x) = [(n + 1)!]^{-1}(d^{n+1}/dx^{n+1})[x^n V^{n+1}(x^{1/2})] \qquad (7.74)$$

Then if we transform $\gamma_n(s)$,

$$W_n(r) = (2/\pi) \int_{r}^{\infty} \gamma_n(s) \, ds/(s^2 - r^2)^{1/2} \qquad (7.75)$$

and use Dirichlet's theorem as in the Firsov procedure, we have

$$W_n(r) = -\int_{r^2}^{\infty} F_n(x) \, dx \qquad (7.76)$$

In the zero-order case, this is very simply

$$W_0(r) = V(r) \qquad (7.77)$$

Thus a knowledge of $\gamma_0(s)$ over a good range of s enables us to establish $V(r)$, drawing from experimental data for as many different energies as we have available.

7.8 INELASTIC COLLISION LOSSES

While we have been treating individual atomic collisions as purely elastic for theoretical purposes, there is undoubtedly some energy loss to the electrons for collision energies in the kilovolt range. In any high energy collision between two atoms the kinetics of the elastic and inelastic effects are rather inextricably entwined. In this chapter so far, however, we have assumed that we may consider the elastic collision separately. We must now ask whether this supposition is justifiable.

The classical energy equations governing the collision will be modified when there is an energy loss to the atomic electrons. With the notation of Section 7.3,

$$E_P = E_P' + E_T + Q \qquad (7.78)$$

$$E_P^{1/2} = E_P'^{1/2} \cos \phi_P + A E_T^{1/2} \cos \phi_T \qquad (7.79)$$

$$E_P'^{1/2} \sin \phi_P = E_T^{1/2} \sin \phi_T \qquad (7.80)$$

It would suffice to measure three of the quantities E_P, E_P', E_T, ϕ_P, and ϕ_T in order to determine Q. We must avoid choosing a combination which necessitates the detection of deflected projectile and target from the same collision, which would require coincidence techniques. This is achieved by measuring the incident projectile energy E_P in connection with quantities connected with either the projectile (E_P' and ϕ_P) or the target (E_T and ϕ_T). From the point of view of determining Q experimentally the latter choice is preferable.

Rearranging Eqs. (7.78)–(7.80) for Q and ϕ_P, we have

$$Q = 2E_T\left\{\left(\frac{AE_P}{E_T}\right)^{1/2} \cos \phi_T - \frac{1 + A}{2A}\right\} \qquad (7.81)$$

and

$$\tan \phi_P = \frac{\sin \phi_T}{(E_P/AE_T)^{1/2} - \cos \phi_T} \qquad (7.82)$$

We note also that if we solve Eq. (7.81) for E_T, we find

$$\frac{E_T}{E_P} = \frac{A}{(1 + A)^2}\left\{\cos \phi_T \pm \left[\cos^2 \phi_T - \frac{1 + A}{A}\frac{Q}{E_P}\right]^{1/2}\right\} \qquad (7.83)$$

Because of the square root there are two possible values of E_T for each ϕ_T and Q. In addition, there is a maximum possible Q for each angle ϕ_T, determined by setting the argument under the square root equal to zero. For angles far from 90°, the target energy corresponding to the negative sign in Eq. (7.83)—the "soft" component—is quite small since Q is not far from the maximum. Then any distribution in Q is more influential in determining the spread of E_T than in the case of the positive sign or "hard" component. The latter will have its energy distributed not far below the elastic target energy.

Morgan and Everhart [14] measured the inelastic energy loss in the manner described above, for collisions of Ar^+ ions on Ar atoms at energies up to 100 keV. Although we shall be more interested in experimental details in the next chapter, it is pertinent here to mention some of the more important points regarding the measurements of angles and energies. Since Q is small compared to E_P and E_T, it is necessary to measure these energies very accurately—any error will be greatly magnified. Similarly a small error in ϕ_T will lead to a large error in Q. Happily this effect decreases as ϕ_T increases, but it is evident that there must be experimental emphasis on accuracy of measurement and precision of incident beam energy.

The most interesting result of Morgan and Everhart was obtained by relating the mean inelastic energy loss $\bar{Q}'$ at various energies to the calculated distance of closest approach R_0 for a collision of the incident ion energy and angle of deflection measured. The latter could be calculated based on a screened Coulomb form of potential which predicted correctly the differential scattering cross sections. A plot of $\bar{Q}'$ against R_0 for energies between 6

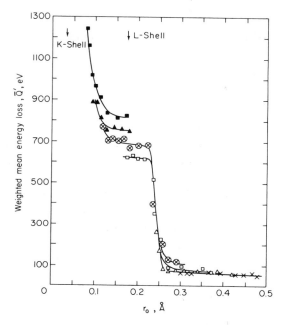

Fig. 7.9. The mean inelastic energy loss $\bar{Q}'$ is plotted against calculated distance of closest approach r_0. Data for 6 different energies are included. Sharp increases are evident at 0.23 and 0.09 Å. The radii of the single argon atom K- and L-shells are indicated: (■) 100 keV, (▲) 75 keV, (⊗) 50 keV, (□) 25 keV, (△) 12 keV, (×) 6 keV. (After Morgan and Everhart [14].)

and 100 keV revealed a universal curve which contained two rapid increases in $\bar{Q}'$, one at about 0.23 Å and another at 0.09 Å (see Fig. 7.9). Therefore, it appears that the inelastic energy loss depends to a markedly greater extent on the interatomic distance than on the incident ion energy. This provides a strong justification for assuming that the inelastic energy loss may be considered independently of the elastic collision. They do of course occur simultaneously in a given collision, which is sometimes termed *quasi-elastic*.

Let us consider such a collision, where the inelastic loss is governed by $Q(R_0)$, such that the relative energy E drops to $E - Q(R_0)$ at the minimum separation (for the elastic collision). Then the scattering angle in the CM system will be given approximately by the mean of those with and without the inelastic loss included, i.e., those starting with relative energy E and $(E - Q)$. This is evidently not strictly correct, since the energy loss to the electrons of the system will not take place suddenly at the distance of closest approach, but will rather occur gradually over a finite part of the trajectory. The results of Morgan and Everhart however lead us to believe that quite a large energy loss occurs over a very small range of separation. It should therefore be a reasonable approximation to assume that the energy disappears between the beginning of the event and the distance of closest approach. Another problem arises in that the inelastic energy loss will inevitably alter slightly the minimum separation R_0 calculated on a purely elastic basis, but this effect is generally assumed to be small. Then the modified deflection and relative energy in the CM system may be used to calculate the laboratory deflections as in Section 7.3.

We have seen in this chapter how experimental results concerning the scattering of atoms or ions may be used to determine the interatomic potential, either directly or through parameter adjustment. In the next chapter we shall describe some types of experiment which provide suitable data.

REFERENCES

1. H. Goldstein, "Classical Mechanics." Addison-Wesley, Reading, Massachusetts, 1956.
2. N. Bohr, *Kgl. Dansk. Vidensk. Selsk. Mat-Fys. Medd.* **18**, No. 8 (1948).
3. Chr. Lehmann and G. Leibfried, *Z. Phys.* **172**, 465 (1963).
4. G. Leibfried and O. S. Oen, *J. Appl. Phys.* **33**, 2257 (1962).
5. C. Lehmann and M. T. Robinson, *Phys. Rev.* **134**, A37 (1964).
6. P. Sigmund and P. Vajda, Risö Rep. Nos. 83, 84 (1964); 115 (1965).
7. E. M. Baroody, *Phys. Fluids* **5**, 925 (1962).
8. J. Lindhard, M. Scharff, and H. E. Schiott, *Kgl. Dansk. Vid. Selsk. Mat.-Fys. Medd.* **33**, No. 14 (1962).
9. M. T. Robinson, *Phil. Mag.* **12**, 741 (1965).
10. O. B. Firsov, *Zh. Esksperim. Teor. Fyz.* **24**, 279 (1953).
11. F. C. Hoyt, *Phys. Rev.* **55**, 664 (1939).
12. G. H. Lane and E. Everhart, *Phys. Rev.* **120**, 2064 (1960).
13. F. T. Smith, R. P. Marchi, and K. G. Dedrick, *Phys. Rev.* **150**, 79 (1966).
14. G. H. Morgan and E. Everhart, *Phys. Rev.* **128**, 667 (1962).

CHAPTER VIII

EXPERIMENTS ON
THE SCATTERING OF ATOMS AND IONS

8.1 Direct Experimental Determination of the Potential

In previous chapters we have discussed numerous examples of the adjustment of parameters contained in a given two-body potential to reproduce theoretically using this potential the results of some form of experiment. Examples are the elastic constants or cohesive energy of a solid or the virial coefficients of a gas. In a sense this procedure might be termed the direct determination of the interaction potential over the range of interatomic separation since the final form of potential hinges on experimental data. But it is always necessary in such estimations to assume a certain form for $V(r)$, based on theory. In this chapter we shall look at some types of experiment which can lead to a direct estimation of the potential without any such prior assumptions.

The scattering of one atom from another is possibly the most unambiguous type of experiment from the point of view of the interatomic potential, although it is technically difficult to perform. The theory behind scattering experiments was outlined in Chapter VII, where we showed how, given the experimental differential scattering cross section, it was possible to use the inversion procedure to derive the interatomic potential. The only assumption was that the $V(r)$ versus r curve contained no maxima or minima. At the present time this is probably the most reliable method available for non-equilibrium potentials, and the results are a good gauge for any theoretically derived form of interaction.

Once the potential curve $V(r)$ is obtained from a scattering experiment over a certain range of r, it is a simple matter to fit a theoretical expression to these results, for analytical treatment of some other atomic collision problem. But care must be taken when extrapolating the fitted potential much beyond the experimental range of interatomic separation, where its validity must fall into question.

8.2   BASIC FEATURES OF ATOM OR ION BEAM EXPERIMENTS

In order to eliminate inelastic scattering phenomena an upper limit to the collision energy must be imposed at a few thousand electron volts. This does not exclude the possibility of potential determinations, since as we have seen in Section 7.8, elastic and inelastic effects may be separated to a reasonable approximation. But if the energy is kept below 3 or 4 keV any inelastic scattering is of negligibly small proportions.

Since it is not technically feasible without elaborate coincidence techniques to record the results of a single collision between two atoms, the quantities measured in scattering experiments are usually the total or differential scattering cross sections S and $\sigma(\phi)$ respectively (see Section 7.5). Essentially, a well-defined, collimated monoenergetic beam of atoms or ions is produced and passed through a scattering chamber containing target atoms at thermal energies. Then the total cross section S may be obtained by measuring the difference in beam intensity reaching a detector placed directly behind the scattering chamber with and without the gas of target atoms present.

Alternatively, measurement of the intensity of beam atoms scattered through a given angle ϕ leads to the differential scattering cross section $\sigma(\phi)$. For direct determination of $V(r)$ using the inversion procedure, it is necessary to measure $\sigma(\phi)$. However, this is rather difficult in practice because of the preponderance of small-angle scattering at greater than thermal energies. Only a small fraction of the incident beam ($\sim 20\%$) undergoes a deflection greater than a few degrees, and of this the percentage scattered into a given solid angle becomes very small. Consequently most early scattering experiments measured the total cross section S and determined the potential by parameter adjustment of an assumed realistic form such as the Bohr or Firsov screened Coulomb interaction or inverse power repulsion. Some differential scattering cross sections for higher energy beams have been measured, however.

Let us consider the theory underlying the interpretation of the results of a beam experiment, neglecting for the moment any practical difficulties. We have typically (see Fig. 8.1) a one-dimensional collimated beam of energy E

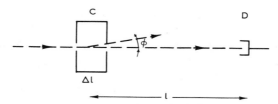

Fig. 8.1. Schematic representation of the basic elements of a scattering apparatus. Collisions occur in the scattering chamber C and undeflected particles are detected at D.

and intensity I_0 scattered by a gas of atoms of density ρ over a distance of beam travel Δl, after which a nonscattered intensity I impinges on a detector D at a distance l from the scattering chamber C. Assuming that there are no multiple scattering events (ρ and Δl small), and that we have a unidimensional detection system (so that the total scattering cross section S is a measure of all particles which do not reach the detector), we have

$$\Delta I/I = -\rho S \,\Delta l \tag{8.1}$$

or by integration

$$I/I_0 = e^{-\rho S \,\Delta l} \tag{8.2}$$

The assumption of a linear detector is somewhat overidealistic from the theoretical point of view, so we suppose that it has an angular aperture of ϕ_0. Then the scattering cross section measured is $S(\phi_0)$ instead of $S(0)$, which following Eq. (7.46), is given by

$$S(\phi_0) = 2\pi \int_{\phi_0}^{\pi} \sigma(\phi) \sin \phi \, d\phi \tag{8.3}$$

This we relate to the interatomic potential using the impact parameter s. From Eq. (7.42)

$$S(\phi_0) = 2\pi \int_{s(\pi)}^{s(\phi_0)} s \, ds = \pi s^2(\phi_0) \tag{8.4}$$

Thus by measuring $S(\phi_0, E)$ at a number of different energies E using Eq. (8.2), the impact parameter for small angle scattering $s(\phi_0, E)$ may be found.

The measurement of $S(\phi_0)$ instead of $S(0)$ is in fact theoretically more useful since the atoms deflected through angles greater than ϕ_0 have been subject to closer collisions than those hitting the detector, which enables us to find the interatomic potential at smaller separations. Varying ϕ as well as altering the energy provides a means of obtaining $V(r)$ over different ranges of r.

The deflection angle $\phi(s, E)$ for a given impact parameter and incident energy is easily found from classical scattering equations if the potential $V(r)$ is known. The theoretical procedure is to choose a physically reasonable form containing adjustable parameters and carry through the calculation of $s(\phi_0, E)$, identifying the parameters by comparison with experiment. In the case of a simple form of potential in the small angle scattering approximation the calculation may be performed analytically, although it is also necessary to assume idealized apparatus geometry. However, it is not difficult to calculate $s(\phi_0, E)$ numerically when either the assumed potential form or the real apparatus geometry require it.

So far we have been considering a highly perfect and unlikely beam experiment which could not be reproduced in the laboratory. The geometry of the apparatus and beam will necessarily influence the interpretation of intensities and angles measured. Thus it is of some importance to estimate the effect of these practical difficulties. First, since the theoretical development hinges on classical mechanics, we must ask to what extent the classical approximation is valid in such an experimental situation. As we have seen in Section 7.1, quantum effects may be neglected even for small-angle scattering for beam energies greater than a few electron volts. The aperture of the detector is much greater than the negligibly small critical angle for classical scattering (see Eq. (7.1)). This is a pleasant result since the inclusion of quantum effects would greatly complicate the theoretical calculation of the scattering cross section.

The experimental difficulties inherent in nonperfect apparatus geometry turn out to be something more of a problem. Reverting to the notation of Fig. 8.1, the length of the scattering chamber Δl is not in any real apparatus negligible compared to the detector distance l. Thus slightly different impact parameters for collisions near the beam entry and exit points will give the same angular deflection ϕ_0 registered at the detector. This effect is slightly increased by some escape of the target gas at the two ends of the chamber. Corrections must be included in the calculation of $S(\phi_0)$ to allow for this.

The geometry of the beam presents another problem. It must have a finite width to provide sufficient intensity at the detector. Then even assuming perfect collimation, that is no beam divergence, some particles may be scattered by the edges of the exit hole of the target chamber. Allowing for these effects renders the determination of $S(\phi_0)$ much more difficult than would appear from Eq. (8.2).

An additional complication, this time theoretical, is introduced if the assumed potential possesses an attractive minimum at low energies. The variation of deflection ϕ with impact parameter s at an incident beam energy E for a potential with an attractive minimum is shown in Fig. 8.2. For this

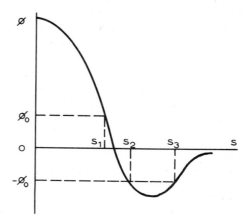

Fig. 8.2. Scattering angles plotted against impact parameter for a potential containing a minimum. At low energies three possible impact parameters s_1, s_2, and s_3 give the same scattering angle ϕ_0.

energy there are three different impact parameters which cause deflections equal to the angle subtended by the detector aperture ϕ_0. If s is less than s_1 or if s lies between s_2 and s_3 the deflected atom will miss the detector. The equation for the scattering cross section becomes

$$S(\phi_0) = \pi[s_1{}^2(\phi_0) + s_3{}^2(\phi_0) - s_2{}^2(\phi_0)] \qquad (8.5)$$

This problem may be averted by increasing the beam energy so that only the repulsive part of the potential influences the scattering. In terms of Fig. 8.2, with increasing energy the minimum of the curve rises until s_2 and s_3 coincide, leading to *rainbow scattering*. For energies greater than this only one impact parameter s_1 contributes to $S(\phi_0)$.

This summarizes the basic essentials of an atomic beam experiment designed to determine the interatomic potential using classical scattering theory. We now consider, again without going into too much detail, some of the apparatus which has been used in these experiments.

8.3 EXPERIMENTAL EQUIPMENT FOR ATOM–ATOM SCATTERING

(1) *Atomic Beam Production*

Since the acceleration of neutral atoms is unfeasible, the general principle is to create ions, accelerate them, and change them back into atoms in flight with no deflection. Several methods of ion beam production are available.

Positive ions may be obtained by a low voltage arc between filament and anode of the ion source and accelerated by an ion gun to the required energy of several kilovolts [1]. Neutralization is accomplished by a small exit hole (∼0.5 mm) in the ion gun, in which some of the ions regain their missing electron by charge transfer without losing energy. Before entering the scattering chamber the remaining ions are removed from the beam by deflection in a lateral field of ∼1000 V produced by a condenser. In a later improvement of this method, Jordan and Amdur [2] used a focused electron discharge to produce the initial ion beam. In this the filament electrons, under the influence of an external solenoid, follow spiral paths through a cylindrical anode. By a system of reflection these electrons are held in the anode cavity sufficiently long to enable the pressure of the atomic gas to be ionized to be held quite low. With this system the pressure could be much lower than that of the low-voltage arc method and still produce the same intensity of electrons. A second improvement is the addition of a differentially pumped charge-transfer chamber containing gas at a pressure which ensures maximum intensity of neutral atoms on leaving. This optimizes the production of neutral atoms after acceleration. Finally, mass and energy analysis of the incident ion beam may be provided by a variable retarding potential or may be achieved by a mass spectrometer interposed between the ion source and charge-transfer chamber.

(2) Scattering Chamber

As has been stated, the dimensions of the target chamber are limited by the requirement that no multiple scattering should occur. Two variables are available for adjustment, namely the length along the beam direction and the gas density. In practice the length is kept fairly small not only so that $\Delta l \ll l$ (see Fig. 8.1), but also to enable a relatively high target gas pressure to be maintained for ease of measurement. Collimation of the beam is provided by the size and configuration of the entry and exit holes of the target chamber. These may be either circular leading to a cylindrical beam or they may take the form of slits, leading to planar geometry. Typical gas pressures in the scattering chamber are of the order of 50 mTorr. Inelastic processes such as ionization occurring during the beam-target collisions are monitored by instruments connected to the scattering chamber. These are however small-probability events at the incident energies of the experiments considered here.

(3) Detector

This is an important item in the apparatus since it is the detector aperture which together with the incident beam width determines the minimum scattering angle ϕ_0 appearing in the total cross section $S(\phi_0)$. The normal type of detector used in atomic beam experiments is a thermal type, either a thermopile or a single fine-wire thermocouple, which is connected through a modulated electronic circuit to an amplifier. Such detectors are insensitive to the miscellaneous neutral rubbish which arrives from the ion source at thermal energies, since the latter produces negligible heating. This is fortunate since such extraneous material constitutes the major portion of the beam, being of considerably greater intensity than that of the fast atoms left after the charge transfer process and residual ion removal.

In the case where the beam has two-dimensional geometry defined by a slit in the target chamber two detectors have been used. One has a receiving width of the same order as the beam width, while the other is a narrow inlet detector used to scan the beam profile.

The results of the atomic beam experiments described above conducted by Amdur and co-workers, are summarized in a very comprehensive review by Mason and Vanderslice [3]. In Table 8.1, we reproduce some values of rare gas atom–atom potentials adjusted to the total scattering cross sections and expressed as inverse power forms.

TABLE 8.1

RARE GAS ATOM–ATOM POTENTIALS[a]

System	$V(r)$ (eV)	Range (Å)	Ref.
He–He	$3.47/r^{5.03}$	0.97–1.48	4
Ne–Ne	$312/r^{9.99}$	1.76–2.13	5
Ar–Ar	$849/r^{8.33}$	2.18–2.69	6
Kr–Kr	$159/r^{5.42}$	2.42–3.14	7
Xe–Xe	$7050/r^{7.97}$	3.01–3.60	8

[a]Obtained by Amdur and co-workers [4–8].

Some recent work by Thompson and co-workers [8a] has extended the Amdur type of experiment to study collisions of a wide range of metal ions of energy 8–25 keV with inert gas atoms. The total cross sections $S(\phi_0)$ were analyzed in terms of inverse power potentials at each incident beam energy, to give experimental points in the potential range between 10 and 100 eV. These were compared with Firsov, Molière, and Csavinsky screened Coulomb potential curves (see Chapters II and III). Agreement was better for the Molière and Csavinsky forms, although the experimental points could in most cases be matched satisfactorily by a simple Born–Mayer potential. Further development of this work may be of considerable interest in clarifying possible shell effects in the dependence of the potential on the atomic number of the accelerated ion (see Section 4.15(4)).

8.4   Measurement of the Differential Scattering Cross Section

Much greater experimental difficulties are encountered in attempting to measure the differential cross section for atoms scattered into a given solid angle. This is mainly because intensities of scattered atoms become rather low, and the thermal type of detector is no longer sufficiently sensitive. Berry [9, 10] has measured the angular distribution of neon and argon atoms in atom–atom scattering into angles 16–64° at energies from 300 to 3000 eV. In his apparatus the scattering chamber contains two detectors, one for the un-deflected beam and another for the scattered beam. These are of the secondary electron ejection type in the form of tantalum electrodes. The off-axis one is shaped to achieve greater sensitivity. Both detectors can be displaced parallel to the beam axis. Thus at a distance x from the entrance slit of the scattering chamber the detector D_2 (see Fig. 8.3) receives atoms scattered through

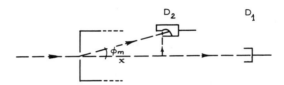

Fig. 8.3. Schematic diagram of the Berry scattering apparatus.

angles between ϕ_m and $\pi/2$. The intensity is therefore greater than that if the cross section were measured directly at a given angle, although the basic geometry is somewhat altered. At different values of x, the incident beam intensity varies because of previous scattering of the target gas. Thus if the detector D_2 is moved back from x to $x + dx$, the intensity at D_2 increases because there are extra scattering centers, but decreases slightly to allow for increased beam attenuation. With the notation of Eq. (8.1)–(8.4), the change in D_2 current is

$$dI = \alpha \rho I_0 [2\pi\sigma(\phi_m) \sin \phi_m \, d\phi_m] \, dx - \rho IS \, dx \qquad (8.6)$$

Here the second term allows for beam attenuation and is obtained by varying the position of D_1 along the beam path. It represents less than 15% of the first term. The coefficient α is a correction to allow for variation in the secondary electron ejection coefficient. By varying x, the differential scattering cross section may be found from this equation.

Berry measured the cross sections for Ne–Ne and Ar–Ar collisions and calculated the interatomic potentials for these two elements by inversion, using procedures described in Section 7.7. He found that the curves thus found could be fitted to within a few percent by forms of exponential potential with constants as in Table 8.2. Applying to a different range of interatomic

TABLE 8.2

ATOM–ATOM POTENTIALS FOR Ne AND Ar[a]

System	$V(r)$ (eV)	Range (Å)	Ref.
Ne–Ne	$6490e^{-4.25r}$	0.4–1.0	10
Ar–Ar	$13700e^{-4.14r}$	0.6–1.2	9, 10

[a]Obtained by Berry [9, 10].

separation than the potentials determined by the Amdur group, the Berry potentials tend to supplement the latter and it is difficult to compare the two sets. At their closest point however they are not incompatible.

8.5 ION–ATOM SCATTERING EXPERIMENTS

The use of ion beams instead of neutral atoms to scatter from a target atom gas has some advantages. Since there is no need to neutralize the accelerated ion beam and consequently lose the nonneutralized part, it is easier to obtain a higher intensity and better collimated beam. Also, after the collision with the target atoms, the deflected beam may be energy-analyzed as well as the different charge states detected by an electric or magnetic transverse field. Increasing the intensity facilitates a measurement of the differential instead of the total cross section, which may then be used to determine the potential directly by inversion. At greater than thermal energies the distinction between ion–ion, ion–atom and atom–atom potentials becomes blurred since the major part of the interaction is closed shell repulsion between the ion cores. Thus for all practical purposes the ion–atom potential found from ion beam experiments is the interatomic potential at energies greater than about 10 eV.

Some early experiments by Simons and co-workers [11] with ion beams concentrated on measuring the total cross section for ion–atom scattering. The main difference between their apparatus and that of the Amdur group is that in the former case the scattering chamber, instead of occupying a very small beam path length, extends as far as the detector entrance (see Fig. 8.4). A small hole in the chamber entrance collimates the beam and a larger one at the combined chamber exit and detector entrance defines the detector aperture ϕ_0. With this geometry $\Delta l \sim l$. This has the advantage that the additional collision path length caused by gas escape from the scattering chamber is negligible compared to the scattering path l. This is however compensated by the fact that the detector aperture for scattered ions depends on where along

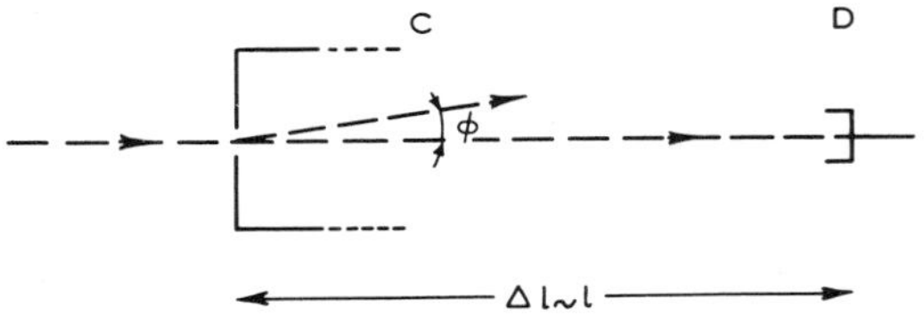

Fig. 8.4. Schematic diagram of the Simons scattering apparatus.

the beam path the collision event occurred, varying between ϕ_0 and $\pi/2$. Consequently an averaging procedure is necessary to obtain the required cross section. This type of geometry was also used in some early experiments by Amdur and Pearlman [12].

In the Simons experiments, the ion beam was accelerated to a higher energy than required, mass-analyzed by magnetic deflection, then retarded to the required energy and focused before passing through a collimating chamber adjacent to the target chamber entrance. In their measurements, an attempt was made to take the charge transfer contribution to the total cross section into account, by detection of the slow ions produced from the atoms in the scattering chamber. This was accomplished by placing a collecting electrode at a small negative potential near the chamber entrance, and the inelastic contribution was then removed from the final measured intensity. It was in itself a rather inaccurate type of measurement, and could not be relied upon to improve the interatomic potential determinations. These were of greatest validity where the probability of charge transfer or other inelastic effects was low.

The ion–atom systems studied by Simons *et al.* were in fact mainly the simple H–He and H–H combinations, although a considerable amount of work was done on ion–molecule systems using both H_2O and organic compounds. Cramer [13, 14] more recently used the Simons technique to study heavier noble gas ion–atom systems. Table 8.3 lists some of the potentials obtained using these techniques.

TABLE 8.3

ION–ATOM POTENTIALS

System	$V(r)$ (eV)	Range (Å)	Ref.
$H_3{}^+$–H_2	$99.8e^{-r/0.376}$	1.48–2.45	11
H^+–H_2	$2.7e^{6(1-r/1.5)}$	1.5–3.7	11
He^+–He	$2.16e^{4.66(1-r/1.08)}$	0.9–3.8	13
Ne^+–Ne	$0.71e^{8(1-r/1.7)}$	1.8–4.2	14

^{a}Obtained by Simons [11] and Cramer [13, 14].

8.6 DEFLECTED BEAM DETECTION EXPERIMENTS

The principal experiments on ion–atom scattering where the deflected ions are detected instead of those passing through with only minor collisions have been performed in the laboratories of Everhart [15–20] and Fedorenko [21–23]. The experimental geometry is illustrated very schematically in Fig. 8.5. The focused beam of accelerated ions (of energy up to 100 keV) enter the scattering chamber containing target gas through a circular aperture. A set of two defining circular orifices, S_1 and S_2, limit the point at which the scattering event occurs to a small volume element about the point P. Different charge states of the scattered beam are then charge–analyzed by electrostatic deflection before passing through into the detector.

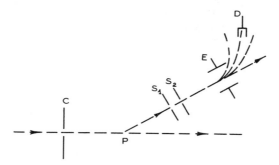

Fig. 8.5. Schematic diagram of the deflected beam detection experiment.

Quite evidently this type of experimental arrangement is very selective and the collision undergone by a particle arriving at the detector may be described with considerable accuracy. Equally evident however is the fact that the scattered intensities measured with respect to the incident beam will be very small. It then becomes essential to have a very efficient and sensitive detection system. The detector used by the Everhart group was a secondary electron multiplier capable of registering currents of incident ions as low as 50 particles per second. There is a secondary consequence of deploying this type of detector, in that the collisions which result in multiply charged scattered ions also produce photons of energies up to a few hundred electron volts. These photons can initiate secondary electrons upon reaching the detector, and produce a current which adds to that produced by the neutral particles passing through the electrostatic analyzer undeflected. It proved rather difficult to separate the currents due to photons and due to neutral particles. The method used was subtraction of the isotropic part of the neutral current (the apparent differential cross section for the neutral component decreased to a constant value at larger angles). This was a factor limiting the scattered angle in the case of He, where most of the scattered particles were neutral, but it did not affect the analysis of results in the case of heavier ion–atom scattering where there were more charged components at larger angles.

Differential cross sections and distribution of different charge states of the scattered particles were measured in the angular range 4–40° with a $\pm 5°$ angular resolution. The incident beam intensity in the target chamber was obtained by a small retractable Faraday cage placed in the path of the beam.

A check of the overall validity of the measurements was provided by comparing the observed differential scattering cross section for He⁺–He collisions with that based on Rutherford scattering. The effect of the atomic electrons in screening the two nuclei of helium was negligible at these energies. Figure 8.6 shows that good agreement was obtained between experimental points and calculated cross sections at incident beam energies 25, 50, and 100 keV, and scattering angles between 4 and 16°.

The experimental curves for differential cross sections for elastic scattering of Ne⁺–Ne and Ar⁺–Ar given by Fuls et al. [18] are shown in Fig. 8.7. The notation is that of the original paper, θ being the scattering angle, b the

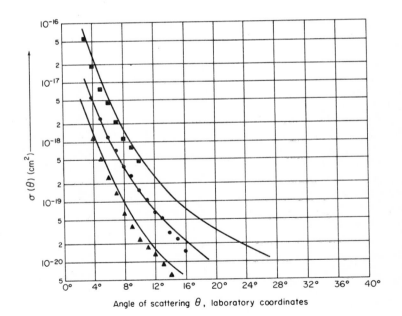

Angle of scattering θ, laboratory coordinates

Fig. 8.6. Differential scattering cross sections for He⁺–He collisions at 25, 50, and 100 keV. The solid lines show cross sections computed on the basis of Rutherford scattering (■) 25 keV, $b/a = 0.016$; (●) 50 keV, $b/a = 0.0078$; (▲) 100 keV, $b/a = 0.0039$. The significance of b/a is explained in the text. (After Fuls et al. [18].)

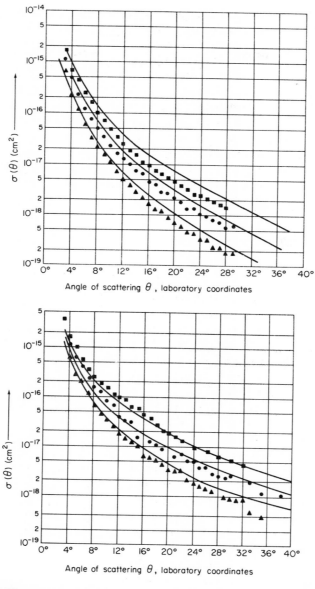

Fig. 8.7. Differential cross sections for (a) Ne$^+$–Ne collisions where (■) 25 keV, $b/a =$ 0.66; (●) 50 keV, $b/a = 0.331$; (▲) 100 keV, $b/a = 0.165$; (b) Ar$^+$–Ar collisions where (■)25 keV, $b/a = 2.63$; (●) 50 keV, $b/a = 1.30$; (▲) 100 keV, $b/a = 0.66$. The solid lines show cross sections computed on the basis of a Bohr potential. The significance of b/a is explained in the text. (After Fuls *et al.* [18].)

collision diameter, and a the Bohr screening length. Experimental cross sections $\sigma(\theta)$ were compared with those calculated on the basis of a Bohr screened Coulomb potential. In almost all cases the experimental points fall consistently slightly below the computed curve, indicating the degree of inaccuracy of the theoretical potential.

As we have noted earlier, experimental data on differential cross sections is susceptible to the Firsov inversion procedure, and may be used directly to find $V(r)$. Results such as those described above were used by Lane and Everhart [15] in just such an inversion calculation. This has been described in Section 7.7, together with the interatomic potentials obtained.

Comparison of the Fuls *et al.* results with those of similar experiments on argon by Kaminker and Fedorenko [21] revealed slight numerical differences but good general consistency.

8.7 MEASUREMENT OF THE INELASTIC ENERGY LOSS

Although we are interested mainly in purely elastic collisions, it is important to examine the nature of the inelastic collisions occurring in the kilovolt energy range to see whether the concept of an elastic potential at these energies has any meaning or application. It is certainly possible to define $V(r)$ from the differential cross section, but we must be able to separate the energy loss through excitation and ionization processes and calculate this in some other way. Unless we can perform this separation, the application of the elastic $V(r)$ to such problems as radiation damage at high energies becomes rather questionable.

Morgan and Everhart [16] used the apparatus described above to perform inelastic energy loss measurements, again in the 25–100 keV energy range for Ar^+–Ar collisions. The theory and conclusions of their experiment have been described in Section 7.8. The main experimental difference was that instead of the scattered incident particles being detected, the energy and angle of the recoil target particles were measured. This was for reasons of convenience in measurement of the inelastic loss Q in a collision. The electrostatic analyzer just before the detector was used to measure the energy of recoil ions in various charge states, after having been calibrated by comparison with the known incident beam energy.

Morgan and Everhart plotted a weighted mean energy loss $\overline{Q'}$ obtained by averaging the energy loss for each charge state, against the recoil angle for a number of incident ion energies between 3 and 100 keV (see Fig. 8.8). The most significant feature of these curves was a sharp increase in $\overline{Q'}$ which occurred at different angles depending on the energy. This led them to investigate the dependence of $\overline{Q'}$ on a theoretically calculated distance of closest approach R_0 based on a screened Coulomb potential. The result was the remarkable set of almost coincident curves reproduced in Fig. 7.9. In effect, the inelastic loss is a strong function of the interatomic distance at all collision energies, having sharp increases at $R_0 = 0.23$ Å and $R_0 = 0.09$ Å for Ar^+–Ar scattering. Thus it is happily a reasonable approximation to separate the inelastic contribution to the energy transfer in an energetic two-atom collision. It may then be calculated by an appropriate method linked to the impact parameter for the collision.

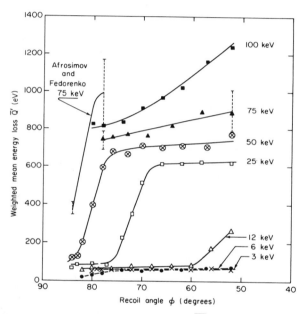

Fig. 8.8. The weighted mean inelastic energy loss $\overline{Q'}$ plotted against the recoil angle ϕ for Ar^+–Ar collisions at a number of different incident ion energies. The 75-keV data of Afrosimov and Fedorenko are also shown. (After Morgan & Everhart [16].)

The data obtained by Morgan and Everhart compared reasonably well with equivalent data at 75 keV given by Afrosimov and Fedorenko [22]. The angular range was rather different since Afrosimov and Fedorenko used a mass spectrometer instead of an electrostatic analyzer and their apparatus was better suited to recoil angles of near 90°. There is a scale discrepancy in the absolute value of $\overline{Q'}$ between the two sets of 75-keV results, but the rapid increase is again evident in the results of the Russian group.

8.8   COINCIDENCE-COUNTING TECHNIQUES IN ION–ATOM COLLISIONS

A step-function increase in the power and adaptability of ion beam experiments is achieved by the use of coincidence counting of the scattered incident particle and recoil particle. Together with energy and charge analyses of the particles, the technique permits a great deal of information to be drawn for individual two-body collisions, with particular emphasis on inelastic processes.

A schematic diagram of the geometry of a coincidence–counting experiment is shown in Fig. 8.9. The incident beam is collimated by holes A_1 and

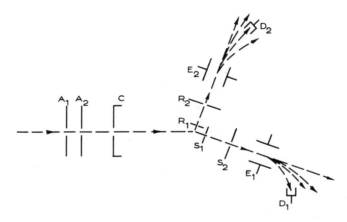

Fig. 8.9. Schematic diagram of a coincidence-counting scattering experiment.

A_2 before entering the scattering chamber through the hole C. Two rotating arms contain the detection systems. The deflected beam is defined by slits S_1 and S_2, and an electrostatic analyzer and detector D_1. The recoil beam has an equivalent system, made up of R_1, R_2, a second analyzer, and D_2. The detectors D_1 and D_2 are connected through an electronic coincidence circuit. The proximity of R_1 and S_1 enable one to locate quite precisely the point at which the collision occurs. Knowing the incident and scattered energies and scattering angles, a classical collision analysis (see Section 7.2) yields the distance of closest approach R_0 and impact parameter s for a given scattering angle θ in the CM system. In this way the missing energy which represents the inelastic contribution Q may be found.

Apparatus based on this geometry has been used by Afrosimov, Gordeev, *et al.* [23] and by Kessel and Everhart [20] to study in detail collisions between argon ions and atoms of the general form: $Ar^+ + Ar \rightarrow Ar^{m+} + Ar^{n+} + (m + n - 1)e$. Although the same in general principle, the two types of apparatus varied in construction. In the experiments of the Russian group, the target gas chamber was small and was connected by metal bellows to the two movable detection arms. The Kessel–Everhart apparatus had the detection arms inside a large target gas chamber and evacuated by separate pumps. Their advantage was that the angular movement was not limited by bellows. Afrosimov *et al.* studied Ar^+–Ar collisions with incident ion energies varying between 12 and 50 keV, while Kessel and Everhart studied collisions from 3 to 400 keV. Above 250 keV the incident ion was Ar^{2+}.

Additional advantages of the coincidence techniques lie in the removal of extraneous counts registering in the detector due to photons from the inelastic collisions or due to incident ion scattering from impurity atoms present in the target gas, which would not register at the same angular coincidence.

In coincidence counting it is important to realize that the idea of "coincidence" is misleading. The scattered and recoil particles do not arrive simultaneously at the detector. Their times of flight depend on the velocities and lengths of flight path, the former of which differs for the two particles except in the case of equipartition of incident energy. The two flight-times may be easily computed from the particle energies and the coincidence gate set to allow for the difference. Another point concerning coincidence counting is that at high counting rates it is necessary to make a statistical allowance for random coincidences of particles from different collision events.

We shall not dwell on the results of the two sets of experiments described above. They contain a large volume of information on multiple charge-transfer events and inelastic collision losses. We would just point out that an extension by Kessel and Everhart of the graphical representation of $\overline{Q'}$ as in the paper by Morgan and Everhart did not contradict the results of the latter. It did appear, however, that there was less tendency for the curves produced by different energy incident beams to lie on the same universal line. Although the dependence on R_0 was still strong, there was an additional variation of $\overline{Q'}$ with the relative velocity of the projectile and target.

8.9 CURRENT AND POSSIBLE FUTURE DEVELOPMENTS IN BEAM SCATTERING

By examining some of the types of experiment which have been used to determine the interatomic potential from collision data, we are in a position to see some of the difficulties involved and perhaps predict some promising future lines of research. The principal problem is, of course, that of intensity. The more accurately we wish to locate the atomic collision producing the particles incident in the detector, the less will be the measured intensity. Thus in measuring the differential scattering cross section instead of the quasi-total cross section, very sensitive multistage secondary electron multipliers must be used. The intensity is even further reduced if a mass-analyser is used in the deflection stage to separate different charge states produced by inelastic collisions. If we then go a stage further, add a second detection system and demand that only the scattered incident particle and target recoil particle from the same collision be registered by coincidence techniques, evidently the intensity with respect to the incident beam intensity is still further reduced.

Assuming that we can achieve sufficient intensity in a very accurately conceived coincidence experiment, other difficulties present themselves. For example, the target atom is not at rest but possesses a thermal momentum perpendicular to the plane of the collision whose instantaneous magnitude is unknown. This will cause it to be scattered out of the plane and the coincidence will be lost. There will not only be a further loss of intensity but a slight inaccuracy in interpretation of the reduced cross sections in terms of the interatomic potential, since the theory generally assumes the target atom to be at rest. The thermal target momentum component parallel to the scattering plane will cause a slight broadening of resolution since coincidence will occur over a small range of scattering angle depending on the relative incident velocity of projectile and target. In the high energy range, these effects are, however, often masked by the angular resolution arising from the finite aperture dimensions of the apparatus, which give natural linewidth effects. Kessel and Everhart [20] noted that thermal effects are a limiting factor only at the low end of the kilovolt range. It is quite naturally a much more significant problem in thermal beam scattering experiments, which are beyond the scope of the present monograph.

The accurate coincidence experiments have so far concerned the scattering of ion beams on target atoms, and sufficient scattered intensity has been achieved to permit measurements of the variation of differential cross section with scattering angle. The logical next step would be to perform similar experiments with atomic beams instead of charged ion beams in order to obtain the true atom–atom elastic potential (although at higher energies this would not be expected to differ greatly from the ion–atom potentials since closed shell repulsion predominates). This step would necessitate an extra stage in the apparatus, similar to that used by Amdur and co-workers, to convert the accelerated ions into atoms by a charge-exchange process. As one can easily see, the problems again multiply. There will immediately be a considerable loss of incident beam intensity. Although the ion–atom conversion is achieved through a resonance process, there will inevitably be a loss of collimation and of energy definition in the resulting atomic beam owing to the elastic part of the charge-exchange interaction. The energy definition may be optimized by a set of fine collimating holes which eliminate all deflected atoms which have lost more than a certain amount of energy defined by the aperture size. But the final result is again a loss of intensity at the target area. Thus an intense and well-defined ion beam before conversion would be an essential for a coincidence atom–atom scattering experiment.

The type of experiment defined in this chapter is difficult to perform and requires much care and precaution. However, it seems at present to be the only good way to study in the laboratory the interaction potential between atoms at energies of more than a few electron volts. The theory largely fails in the intermediate energy range and we must resort to empirical forms of interaction to describe atomic collisions occurring, for example, in radiation damage events in solids. Scattering events provide a valuable check on the validity of such potentials, so further development of laboratory atomic collision studies, in particular, an extension to the case of metallic atoms, would be extremely desirable.

REFERENCES

1. I. Amdur and A. L. Harkness, *J. Chem. Phys.* **22**, 664 (1954).
2. J. E. Jordan and I. Amdur, *J. Chem. Phys.* **46**, 165 (1967).
3. E. A. Mason and J. T. Vanderslice, "Atomic and Molecular Processes" (D. R. Bates, ed.), p. 663. Academic Press, New York, (1962).
4. I. Amdur, J. E. Jordan, and S. O. Colgate, *J. Chem. Phys.* **34**, 1525 (1961).
5. I. Amdur and E. A. Mason, *J. Chem. Phys.* **23**, 415 (1955).
6. I. Amdur and E. A. Mason, *J. Chem. Phys.* **22**, 670 (1954).
7. I. Amdur and E. A. Mason, *J. Chem. Phys.* **23**, 2268 (1955).
8. I. Amdur and E. A. Mason, *J. Chem. Phys.* **25**, 624 (1956).
8a. C. Foster. I. H. Wilson, and M. W. Thompson. *J. Phys. B*, to be published (1972).
9. H. W. Berry, *Phys. Rev.* **75**, 913 (1949).
10. H. W. Berry, *Phys. Rev.* **99**, 553 (1955).
11. J. H. Simons, C. M. Fontana, E. E. Muschlitz, and S. R. Jackson, *J. Chem. Phys.* **11**, 307 (1943).
12. I. Amdur and H. Pearlman, *J. Chem. Phys.* **8**, 7 (1940).
13. W. H. Cramer and J. H. Simons, *J. Chem. Phys.* **26**, 1272 (1957).
14. W. H. Cramer, *J. Chem. Phys.* **28**, 688 (1958).
15. G. H. Lane and E. Everhart, *Phys. Rev.* **120**, 2064 (1960).
16. G. H. Morgan and E. Everhart, *Phys. Rev.* **128**, 667 (1962).
17. E. Everhart, G. Stone, and R. Carbone, *Phys. Rev.* **99**, 1287 (1955).
18. E. W. Fuls, P. R. Jones, F. P. Ziemba, and E. Everhart, *Phys. Rev.* **107**, 704 (1957).
19. G. H. Lane and E. Everhart, *Phys. Rev.* **117**, 920 (1960).
20. Q. C. Kessel & E. Everhart, *Phys. Rev.* **146**, 16 (1966).
21. D. M. Kaminker and N. V. Fedorenko, *Zh. Tekh. Fiz.* **25**, 2239 (1955).
22. V. V. Afrosimov and N. V. Fedorenko, *Zh. Tekh. Fyz.* **27**, 2557 (1957) [English transl.: *Sov. Phys. Tech. Phys.* **2**, 2391 (1957)].
23. V. V. Afrosimov, Yu. S. Gordeev, M. N. Panov, and N. V. Fedorenko, *Zh. Tekh. Fyz.* **34**, 1613, 1624, 1637 (1964) [English transl.: *Sov. Phys. Tech. Phys.* **9**, 1248, 1256, 1265 (1965)].

CHAPTER IX

LIQUID METAL PAIR
INTERACTION POTENTIALS

9.1 LIQUID METALS

The exploration of the liquid state of metals has developed extremely rapidly during the past few years, and a great deal of information is accumulating about their structure and physical properties. It is a well-known fact that the volume change of a metal upon melting is sufficiently small to permit extrapolation of information on interatomic forces from liquid to solid and vice versa. The nature of these forces is not as yet fully established, but the problem is most commonly approached in the pair approximation. The atoms in the liquid are assumed to interact in pairs and three-body or higher-order forces are neglected. Physical properties are then computed on this basis. Thus liquid metal data, interpreted on the basis of pair interactions, provide a useful tool for probing the interatomic pair potential at near-equilibrium energies. A considerable amount of work has recently been carried out towards obtaining a pair potential from such data, with results which are interesting although (to date) not quite conclusive.

In this chapter we shall describe in an elementary manner the relationship between the properties of a simple liquid metal and the interatomic potential, and attempt to present a survey of recent work in the field. At the end of the chapter we shall try to indicate some paths along which liquid metal research is likely to develop. Considerable risk is however attached to any prediction in a field which is in such a rapid state of flux. For more details of various aspects of the subject, readers are referred to two recent books on liquid metals [1, 2] and to original references, in particular those contained in the proceedings of an international conference which covered the field quite thoroughly [3].

9.2 CORRELATION FUNCTIONS AND ATOMIC INTERACTIONS

The forces between atoms in a liquid are related to physically measurable quantities concerning the structure of the liquid through the medium of statistical mechanics. The fundamental step is to define a set of distribution functions of the type n_i such that

$$n_i(\mathbf{r}_1, \mathbf{r}_2, \mathbf{r}_3, \ldots, \mathbf{r}_i)\, d\mathbf{r}_1\, d\mathbf{r}_2\, d\mathbf{r}_3 \cdots d\mathbf{r}_i \qquad (9.1)$$

represents the probability of finding the ith atom in a volume element at $\mathbf{r}_i$ if atoms 1 to $(i-1)$ are all situated in their respective volume elements at $d\mathbf{r}_1$ to $d\mathbf{r}_{i-1}$. Of these the simplest (apart from the single atom distribution function $n_1(\mathbf{r}_1)$) is the two-atom distribution function $n_2(\mathbf{r}_1, \mathbf{r}_2)$. This is more commonly replaced by a *radial distribution function* or *pair correlation function* $g(r)$, since it is independent of $\mathbf{r}_1$ if the liquid is in thermal equilibrium, depending only on the distance r between the two atoms. To define $g(r)$ the first atom is considered to be at the origin and the probability is normalized to unity at large r:

$$g(r) = n_2(0, \mathbf{r}_2)/\rho_0{}^2 \qquad (9.2)$$

where ρ_0 is the average density of the liquid.

In order to obtain a relation between the statistical liquid structure and interatomic forces we shall require the *triplet correlation function* of the form

$$n_3(\mathbf{r}_1, \mathbf{r}_2, \mathbf{r}_3)\, d\mathbf{r}_1\, d\mathbf{r}_2\, d\mathbf{r}_3 \qquad (9.3)$$

Its definition follows from that of the general function of Eq. (9.1).

If we now assume that the liquid may be described in terms of pair potentials $V(r)$ between atoms, a simple analysis of a three-atom configuration leads to an expression involving this potential. We express $g(r)$ in terms of a potential of mean force $U(r)$ using standard statistical mechanics:

$$g(r) = e^{-U(r)/kT} \tag{9.4}$$

Consider two atoms placed at $\mathbf{r}_1$ and $\mathbf{r}_2$ respectively (Fig. (9.1)) The total

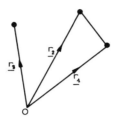

Fig. 9.1. A three-particle distribution referred to an origin O. (Underlined letters correspond to boldface in text.)

force acting on atom 1, which is just the negative derivative of the potential of mean force, may be divided up into that due to atom 2 and that due to all the remaining atoms:

$$-dU(r)/d\mathbf{r}_1 = -dV(r)/d\mathbf{r}_1 + \sum_i F_i \tag{9.5}$$

where $r = |\mathbf{r}_2 - \mathbf{r}_1|$.

To find $\sum F_i$ we remember that the probability of finding a third atom at $\mathbf{r}_3$ having already fixed atoms 1 and 2 is

$$n_3(\mathbf{r}_1, \mathbf{r}_2, \mathbf{r}_3)\, dr_1\, dr_2\, dr_3 = \frac{n_3(\mathbf{r}_1, \mathbf{r}_2, \mathbf{r}_3)\, d\mathbf{r}_3}{n_2(\mathbf{r}_1, \mathbf{r}_2)}$$

$$= \frac{n_3(\mathbf{r}_1, \mathbf{r}_2, \mathbf{r}_3)\, d\mathbf{r}_3}{\rho_0{}^2 g(r)} \tag{9.6}$$

Therefore the force on atom 1 as a result of atom 3 is

$$F_3 = -\frac{n_3(\mathbf{r}_1, \mathbf{r}_2, \mathbf{r}_3)}{\rho_0{}^2 g(r)} \cdot \frac{\partial V(r_{13})}{\partial \mathbf{r}_1} \tag{9.7}$$

with $\mathbf{r}_{13} = |\mathbf{r}_3 - \mathbf{r}_1|$.

So Eq. (9.5) becomes finally, upon changing the summation to an integral,

$$\frac{dU(r)}{d\mathbf{r}_1} = \frac{dV(r)}{d\mathbf{r}_1} + \int \frac{n_3(\mathbf{r}_1, \mathbf{r}_2, \mathbf{r}_3)}{\rho_0{}^2 g(r)} \frac{\partial V(r_{13})}{\partial \mathbf{r}_1}\, d\mathbf{r}_3 \tag{9.8}$$

If we can measure $g(r)$ experimentally then Eqs. (9.4) and (9.8) may be used to find $V(\mathbf{r})$, the only obstacle being the triplet correlation function n_3, about which very little is known, although it is attracting an increasing amount of experimental interest. In current theories n_3 is replaced by approximate estimates, as in the superposition approximation of Kirkwood, in which it is expressed as a product of pair correlation functions

$$n_3(\mathbf{r}_1, \mathbf{r}_2, \mathbf{r}_3) = \rho_0{}^3 g(r_{12}) g(r_{23}) g(r_{31}) \qquad (9.9)$$

We shall examine the results of such approximations in Section 9.4. First let us look at the relation between correlation functions and the liquid metal structure factor since this is an important quantity in experimental methods of determining liquid structure using X-ray, electron, or neutron scattering.

9.3 THE LIQUID METAL STRUCTURE FACTOR

The structure factor $S(q)$, which fulfills a function similar to that of the crystal lattice structure factor of Chapter V, is proportional to the scattering cross section in the liquid. It is related to the pair correlation function $g(r)$ by the equation

$$S(q) = 1 + \rho_0 \int [g(r) - 1] e^{\mathbf{q} \cdot \mathbf{r}} \, d\mathbf{r} \qquad (9.10)$$

Since the liquid is rotationally invariant it may be expressed in polar coordinates:

$$S(q) = 1 + 4\pi\rho_0 \int_0^\infty r^2 [g(r) - 1] \frac{\sin qr}{qr} \, dr \qquad (9.11)$$

This equation would give us the scattered intensity of radiation if we knew the function $g(r)$. However, since $S(q)$ is more experimentally accessible than $g(r)$ it is more useful to perform a Fourier inversion of Eq. (9.10) or (9.11) to obtain $g(r)$ in terms of $S(q)$, which can then be obtained from experiment. Thus

$$g(r) = 1 + \frac{1}{(2\pi)^3 \rho_0} \int [S(q) - 1] e^{\mathbf{q} \cdot \mathbf{r}} \, dq \qquad (9.12)$$

$$= 1 + \frac{1}{2\pi^2 r \rho_0} \int_0^\infty [S(q) - 1] q \sin qr \, dq \qquad (9.13)$$

In terms of a *total correlation function* $h(r)$, defined by

$$h(r) = g(r) - 1 \qquad (9.14)$$

we have

$$h(r) = (1/2\pi^2 r \rho_0) \int_0^\infty [S(q) - 1] q \sin qr \, dq \qquad (9.15)$$

and

$$h(q) = S(q) - 1 \qquad (9.16)$$

A further type of correlation function which is especially applicable to pair interactions is the Ornstein–Zernike *direct correlation function* $c(r)$, obtained by splitting $h(r)$ into an effect due to a two-atom interaction plus another due to the remaining atoms, as in the case of the potential

$$h(r) = c(r) + \rho_0 \int c(|\mathbf{r} - \mathbf{r}'|)h(\mathbf{r}')\, d\mathbf{r}' \tag{9.17}$$

By Fourier transformation it is easily shown that

$$c(r) = (1/2\pi^2 r\rho_0) \int_0^\infty [1 - 1/S(q)]q \sin qr\, dq \tag{9.18}$$

and

$$c(q) = 1 - 1/S(q) \tag{9.19}$$

It is not possible to measure the structure factor for very small angle scattering ($q \sim 0$), but there is a standard thermodynamical result which may be applied to the long wavelength limit:

$$S(0) = \rho_0 kTK_T \tag{9.20}$$

where K_T is the isothermal compressibility.

Notwithstanding this theoretically available point on the curve, it turns out to be of vital importance for potential determination to measure the structure factor very precisely to as low a value of q as possible. The long range part of the potential is very sensitive to this section of the $S(q)$ curve. This will be demonstrated in a later section. The direct correlation function, through its transform $c(q)$, is useful in that it accentuates the numerically small parts of $S(q)$.

9.4   SOLUTIONS FOR THE PAIR POTENTIAL

There are two principal theories used in the solution of Eq. (9.8) for $V(r)$—the so-called Born–Green [4] (BG) and Percus–Yevick [5] theories (PY)—which we shall now describe.

The BG theory makes use of the superposition approximation to describe n_3 (see Eq. (9.9)). By substituting this into Eq. (9.5) and integrating, the latter may be expressed in a form analogous to that defining the direct correlation function (see Eq. (9.17)):

$$U(r)/kT = (V(r)/kT) - \rho_0 \int E(\mathbf{r} - \mathbf{r}')h(\mathbf{r}')\, d\mathbf{r}' \tag{9.21}$$

where

$$E(r) = \int_r^\infty (V'(x)/kT)g(x)\, dx \tag{9.22}$$

The large-r limit of $E(r)$ is easily seen to be $-V(r)/kT$ since $g(x) \to 1$, and from Eqs. (9.4) and (9.14), $h(r)$ approaches $-U(r)/kT$ at large r. Thus if we compare Eqs. (9.17) and (9.21) for large values of r, we can see an equivalence between $c(r)$ and $E(r)$. Hence there is an intimate connection between $c(r)$ and $V(r)$ similar to that existing between $h(r)$ and $U(r)$, except of course that the former is restricted to an asymptotic relationship. Based on this analogy we may write as an approximation for $c(r)$,

$$c(r) \simeq e^{-V(r)/kT} - 1 \tag{9.23}$$

Since this is already under the restriction of large r, we can go one step further and write

$$V(r) \simeq -kTc \tag{9.24}$$

This offers a crude method of finding an asymptotic pair potential from the measured structure factor via the direct correlation function. At large r, $c(r)$ should be the negative of the pair potential in units of kT.

Slightly increased accuracy is achieved by the so-called *hyperchain* theory, which assumes that $E(r)$ may be replaced by $c(r)$ at all values of r, so that

$$U(r)/kT = V(r)/kT - \rho_0 \int c(\mathbf{r} - \mathbf{r}')h(\mathbf{r}')\, d\mathbf{r}' \tag{9.25}$$

Then using Eq. (9.17), we have

$$\ln[1 + h(r)] = -(V(r)/kT) + h(r) - c(r) \tag{9.26}$$

or

$$V(r) = kT\{h - c - \ln(1 + h)\} \tag{9.27}$$

In the PY theory, which resembles the hyperchain theory, an equation similar to (9.26) is obtained, with the factor $[h(r) - c(r)]$ replaced by $\ln[1 + h(r) - c(r)]$:

$$\ln[1 + h(r) - c(r)] = \ln[1 + h(r)] + V(r)/kT \tag{9.28}$$

or written in terms of $g(r)$,

$$V(r) = kT \ln(1 - c/g) \tag{9.29}$$

We see, therefore, that if we possess experimental information on the structure of a liquid metal and wish to obtain a pair potential, we have a certain amount of choice regarding the method, although all three approximations are restricted to low-density liquids. The BG equations (9.21) and (9.22) may be solved directly by numerical methods using experimental data for the correlation functions involved. Alternatively we may use the hyperchain or PY equations (9.27) and (9.29) to obtain more rapid solutions applicable to the asymptotic region of large r, and consequently restricted to extremely low densities. Superior to all of these methods, of course, would be the direct solution of the exact equation (9.8) if the three-atom correlation function were available.

9.5   INTERATOMIC POTENTIALS OBTAINED
FROM LIQUID METAL DATA

Having established the accurate and approximate relations between the structural features of a liquid metal and the pair potential, let us examine some potentials derived from experiments in the liquid state using these theories.

Johnson et al. [6] used both BG and PY theories to calculate pair interactions for eight metals (the alkalis plus Hg, Al, and Pb) based on distribution functions obtained from X-ray and neutron scattering data. In all cases their potential contained long-range oscillations following a deep and relatively narrow minimum. Their BG potential for Na is shown in Fig. 9.2. The PY potential was found to be rather more dependent on the temperature at which the scattering experiments were performed, and consequently it was assumed that the BG approximation was more physical. When the analysis of Johnson et al. (JHM) was applied to Ar a potential containing a single minimum and monotonically increasing tail ensued in good agreement with rare gas theory. This they interpreted as confirming the validity of their method of finding the liquid metal potentials and of the oscillatory nature of these interactions at large r. More recent work has however thrown the JHM potentials into question, as will be indicated later.

A rather different form of metal potential was obtained by Ascarelli [11] from PY theory, working directly from q-space data. This is illustrated by Fig. 9.3 showing the Ascarelli pair potential for Ga. There is a pronounced first maximum followed by an extremely irregular and temperature-sensitive curve. The amplitude of the first maximum in the potential appears to be rather sensitive to the small-q behavior of the structure factor $S(q)$, rather than solely a feature of the discontinuity at the Fermi surface. We shall consider this sensitivity in greater detail in the next section.

North et al. [12] obtained potentials for liquid lead which slightly resembled those of Ascarelli, using both PY and hyperchain theories. The work was recently extended by Gehlen and Enderby [13] who solved the BG

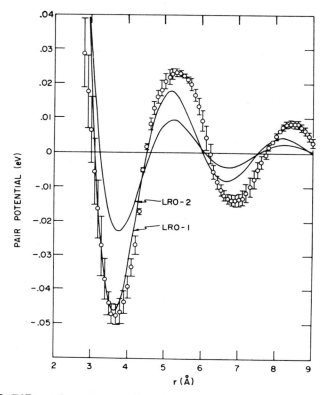

Fig. 9.2. Different long-range oscillatory potentials applied to liquid sodium. The experimental points are those of JHM [6] obtained using BG theory and X-ray data [8]. The potential LRO-1 is the PR [9] potential fitted to these points. LRO-2 has a shallower minimum and reproduces better the neutron data of Gingrich and Heaton [10]. (After Paskin [7].)

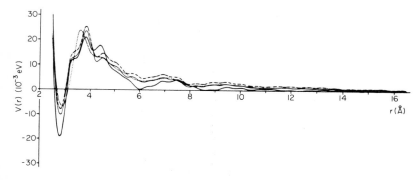

Fig. 9.3. Pair potential obtained by Ascarelli for Ga using PY theory at varying temperature and pressure. (After Ascarelli [11].)

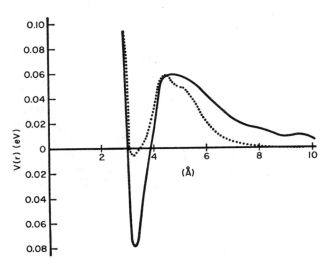

Fig. 9.4. BG (solid line) and PY (dotted line) pair potentials for liquid lead at 780° C obtained by Gehlen and Enderby. (After Gehlen and Enderby [13].)

equation in q-space. A comparison of the pair potentials from BG and PY theory obtained for Pb is shown in Fig. 9.4. Gehlen and Enderby demonstrated that the effect on $V(r)$ of varying the structure factor for low q over the experimental error range was large in the region of r between 5 and 15 Å. They thereby confirmed the necessity for more careful measurements of $S(q)$ for small-angle scattering.

9.6   SENSITIVITY OF THE PAIR POTENTIAL TO LIQUID STRUCTURE

From the theoretical considerations of this chapter, the way seemed to be clear for a quite direct and unambiguous determination of the interatomic potential from liquid metal structure factors and correlation functions. Unfortunately, this desirable state of affairs was marred by an investigation of the sensitivity of the main features of liquid structure to details of the pair interaction.

The interaction between atoms in a solid or liquid metal may with varying accuracy be described by potential forms varying from a hard-sphere model through a Lennard-Jones inverse power form to a long-range oscillatory ion–electron–ion interaction. Ashcroft and Leckner [14] used a hard-sphere potential with variable radius to calculate the structure factors of a number of metals in the PY approximation. The resulting comparison with the experimental $S(q)$ for K and Rb is shown in Fig. 9.5. Considering the inadequacy of

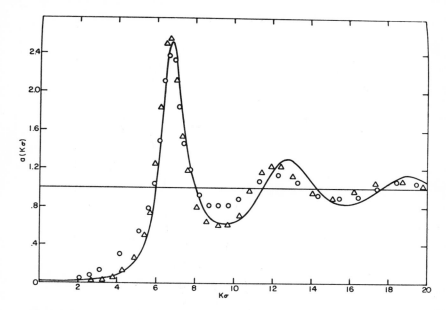

Fig. 9.5. Structure factors for K and Rb from neutron diffraction data [10], compared with the PY hard-sphere theory (solid line) for a fluid of packing density 45%, computed by Ashcroft and Leckner. The notation is that of the original reference; a is the structure factor, σ the ionic radius, and K the wave number: (○) K, 65°C, $\sigma = 4.07$ Å; (△) Rb, 40°C, $\sigma = 4.30$ Å. (After Ashcroft and Leckner [14].)

the potential, the agreement is quite striking, and illustrates quite clearly that it is the interaction between ion cores which decides the main structural features of $S(q)$ and $g(r)$. We must therefore look into the detailed behavior of $S(q)$ to find any effect on the long-range part of $V(r)$.

It has been generally assumed that some long-range oscillations due to imperfect screening of the ion cores by the conduction electrons should be present in the metal potential (see Chapter V). Experimental confirmation of this appeared to be forthcoming in the pair interaction obtained by JHM from the measured radial distribution function.

The computer simulation of molecular dynamics (see Section 9.7) provides a method of checking the consistency of various potentials, since any two-body interaction may be used to provide a resulting radial distribution function. Paskin and Rahman [7, 9] compared the radial density function $4\pi \bar{r}^2 \rho_0 g(r)$ for Na obtained from a potential containing long range oscillations and closely resembling the JHM potential, with the experimental curve from X-ray scattering data [8]. They found that the general features of the experimental curve were reproduced by this potential (see Fig. 9.6). However, what was more significant was the result of carrying out the molecular

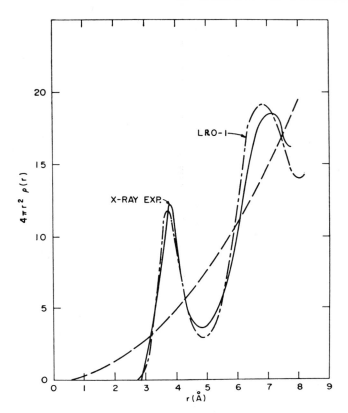

Fig. 9.6. Comparison of the radial density functions for Na obtained from X-ray data [8] and from molecular dynamic computations [9] using the potential LRO-1. (After Paskin [7].)

dynamic computations using a Lennard-Jones potential with suitable parameters. There was very little difference between the Lennard-Jones radial density function and that obtained from their long-range oscillatory potential although the two potentials differed greatly in the region beyond the first minimum (see Fig. 9.7).

Further confirmation of this insensitivity of structure to potential is provided by a comparison by Paskin [7] of the experimental $g(r)$ for Ar with those for several liquid metals, scaled so that in all cases the position of the first maximum coincides (see Fig. 9.8). This scaling corresponds to a division of the r-scale by a factor proportional to atom size, resulting in a striking coincidence of the correlation functions.

The conclusion of these comparisons is that the easily measurable aspects of liquid structure are dependent only on the depth and position of the first

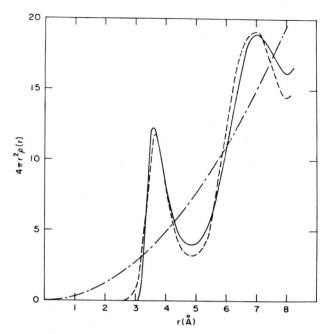

Fig. 9.7. Comparison of the radial density functions obtained by PR [9] for a Lennard-Jones potential, as opposed to a long-range oscillatory potential for liquid Na near the melting point: (——) Lennard-Jones; (– – –) LRO-1; (— · —) average density. (After Paskin [7].)

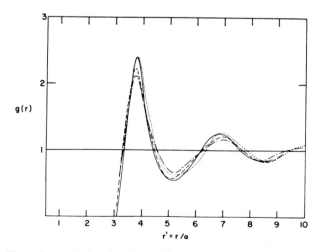

Fig. 9.8. The pair correlation functions $g(r)$ of liquid metals and liquid Ar, scaled so that the position of the first maximum coincides :(- - -) Ar (n), $a = 1$; (——) Cs (n), $a = 1.39$; (— · —) Na (n), $a = 1$; (– – –) Rb (n), $a = 1.3$. The magnitude of $g(r)$ is not scaled. (After Paskin [7].)

potential minimum. They are influenced to only a very minor degree by the detailed shape of the potential curve, especially in the region beyond this minimum. Paskin [7] observed that even the crude model for a liquid which involves the random packing of spheres gives reasonable agreement with experiment for the radial distribution function. The validity of the JHM oscillatory potentials is therefore extremely dubious, and confirmation of the presence of long-range oscillations must await a very precise experimental determination of liquid structure especially at large separations (small angle scattering). It would also be desirable to refine the theory connecting the liquid correlation function and the interatomic potential, perhaps using a better approximation for the triplet correlation function.

9.7   The Computer Experiment Applied to Liquid Metals

A technique which has been adapted to complement the transition from distribution function to pair potential is that of the computer experiment in molecular dynamics. Rahman [15] simulated on a large capacity computer a system of about 800 atoms in a box with periodic boundary conditions. With these atoms interacting under the influence of a pair potential it was possible to obtain information about the structure of such a system. Solving the classical equations of motion for the ensemble yielded a pseudoexperimental radial distribution function as well as a time-dependent mean square displacement of the atoms, which in turn led to a coefficient of self-diffusion. First of all the simulation technique was applied to Ar using a Lennard-Jones potential. Both the experimental radial distribution function and the self-diffusion coefficients were reproduced by the computer results.

This computer technique, though limited by the number of atoms and volume size which may be simulated, is a valuable aid in attacking the liquid metal structure-potential relationship. In collaboration with Paskin, Rahman [9] used two different Na potentials containing long-range oscillations (LRO-1 and LRO-2) as the basic pair interactions in their simulated liquid. These potentials, included in Fig. 9.2, were both truncated at 8.18 Å. LRO-1 was close to the JHM potential and LRO-2 had the same theoretical form (a combination of BM and $A \cos(ar)/r^3$ forms) but a shallower first minimum close to that of the Cochran potential (see Section 6.9). They concluded that LRO-2 reproduced more accurately than LRO-1 the experimental radial distribution function from which the JHM potential had been obtained, and consequently that the BG and PY approximations were inadequate. Unfortunately Paskin and Rahman had been misled by an error contained in the JHM paper, in which the authors stated that their potential had been derived from neutron scattering data of Gingrich and Heaton [10]. In actual fact it was the X-ray results of Orton, Shaw, and Williams [8] which were used by JHM. As Fig. 9.9 demonstrates, there is a considerable discrepancy between the two sets of experimental data for Na, and in fact the Paskin–Rahman LRO-1 potential yields a radial density function in good agreement with the X-ray data. There is therefore consistency between experiment and theory and the approximations involved in the theory have been cleared of major suspicion.

A further investigation of the sensitivity of the structure factors obtained by computer experiments to the assumed pair potential has been carried out by Schiff [16]. In particular, it was asked whether and under what conditions

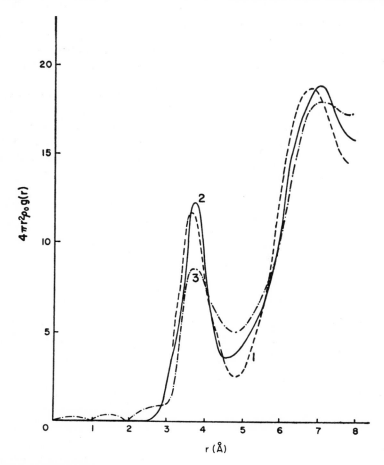

Fig. 9.9. Radial density function for Na obtained by PR [9] using the potential LRO-1 (1) compared with that from X-ray data (2) and neutron data (3). (After March [1].)

the presence of long-range oscillations or the hardness of the ion core would produce measurable features in the structure factor. Schiff compared two ion core potentials—a soft one given by a BM form and a hard one represented by a $1/r^{12}$ term. The effect was intuitively reasonable. For the softer potential the structure factor oscillations were more damped (see Fig. 9.10). However, the difference in the two $S(q)$ was negligible for the first maximum and only became evident at larger values of q. An accurate determination of $S(q)$ at large q would therefore be essential in order to obtain information regarding the steepness of the core repulsion.

To test the effect of long-range oscillations, Schiff used three models— the asymptotic form due to Pick [17], the LRO-2 potential of Paskin and Rahman [9], and another containing still larger oscillations. The effect on the amplitude of the structure factor $S(q)$ was noticeable only for unreasonably large oscillations in $V(r)$. This suggests that it would be very difficult to confirm the presence or absence of long-range oscillations in the pair potential from experimental structure data.

A more sensitive method of detecting such oscillations again proposed by Schiff was to vary their wavelength by mixing two liquid metals such as Li and Mg, of nearly equal hard core diameters and different valence. In a simple model using an asymptotic form of oscillatory potential, the height of the first peak of the structure factor $S(q_0)$ has a maximum when $2k_F = q_0$.

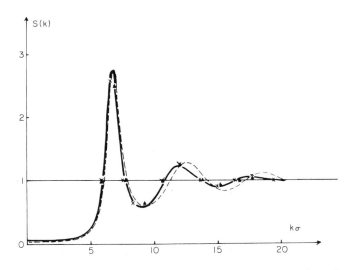

Fig. 9.10. Structure factors for Na obtained by Schiff in molecular dynamic computations, for BM (solid line) and Lennard-Jones (dashed line) ion cores. The crosses and triangles denote experimental points. The unit of length o is an approximate hard core radius of 3.24 Å, and k is equivalent to q in the text. (After Schiff [16].)

Variation of k_F is obtained by varying the percentages of the constituents of the liquid Li–Mg alloy. If this resonance were detected by experiment it could provide some information on the long-range oscillations in $V(r)$.

It is extremely promising to find that experimental structure data for liquids may be interpreted in terms of a pair interaction between the atoms. Nevertheless, as has been demonstrated by computer experiments and analytical considerations, it is essential to investigate the sensitivity of the potentials obtained to both the precision of experimental results and the approximations involved in the theory. Indeed, the major difference in form between the JHM and the Ascarelli or Gehlen–Enderby types of potential would indicate that further and more precise work is necessary.

Considering the accuracy presently available in experimental structure determinations, it might be better to concentrate on the dynamical aspects of computer experiments such as velocity autocorrelation functions and self-diffusion coefficients obtained from different interatomic potentials. These appear to be more sensitive to the pair interaction. Paskin and Rahman found that at $100°C$ using their LRO-1 form the self-diffusion coefficient D was 1.9×10^{-5} cm^2/sec, while for LRO-2 D was 5.8×10^{-5} cm^2/sec, as compared to an experimental value of 4.2×10^{-5} cm^2/sec. Schiff demonstrated that the velocity autocorrelation function was quite sensitive to the core steepness of the potential. The time-dependent properties of liquids may therefore prove more fruitful than the time-averaged structural aspects from the point of view of potential determination.

9.8 Future Developments in Liquid Metal Potentials

A great deal of the work described in this chapter has been initiated during the past five years. It is therefore a preliminary report of a quite promising field of study which has been presented. As such it must inevitably be over-taken very rapidly by further progress in the liquid metal field. It would perhaps be useful to summarize briefly the present situation from the view-point of the interatomic potential.

The principal theoretical difficulty lies in the lack of information on the triplet correlation function n_3. If this were known, an exact relation between structure and pair potential could be established (see Eq. (9.8)). In the absence of such information it appears that the superposition approximation (see Eq. (9.9)) is the best, although still doubtful at liquid metal densities. This then permits a solution of Eq. (9.8) to varying degrees of approximation, as described in Section 9.4.

Experimental work on potential determination coupled with computer simulation of molecular dynamics shows quite clearly that the main aspects of liquid structure are decidedly insensitive to details of the pair interaction. Potentials with and without long-range oscillations reproduce the experimental correlation functions within present experimental accuracy. Conversely, the finer features of the potential are extremely sensitive to the structure factor measurements, which means that a high degree of experimental precision will be required to achieve a reliable potential. The existence or importance of a positive region of potential beyond the first minimum, without which metallic and nonmetallic liquids are equivalent, cannot be said to have been established beyond doubt.

From the results so far we are forced to the conclusion that it would be unrealistic to place a great deal of confidence in any of the pair interactions which have been presented by a variety of authors in the field of liquid metals. These potentials have an enormous degree of variation with the theory used to relate structure factor and pair interaction, and with the experimental temperature. They vary from highly perfect Friedel oscillations to extremely complicated and irregular curves.

To search for the reason for this rather unsatisfactory state of affairs, we must examine the validity of the basic theoretical assumptions in relation to the experimental conditions. A liquid metal is in fact a relatively dense medium, so its representation by two-body interactions is rather doubtful. This is especially so for scattering at small values of q, where there are annoying effects due to the electron gas. We have mentioned that it is this region which is important in determining the potential at large r. Thus it may not even suffice to have precise measurements of small angle scattering in order to find the potential, in particular if BG, PY or hyperchain approximations are used to provide the link between structure and potential. Such approximations are really at the outside limit of their validity at liquid metal densities, and a better method of obtaining the triplet correlation function would be desirable.

While the potential is insensitive to the averaged liquid properties, it appears to be more accessible to the time-dependent properties such as diffusion. Time-dependent correlations can be developed and related to scattering data (see for example March [1]). Again, such properties are relatively easily obtained from the computerized molecular dynamics approach. This appears to be a promising future line of development.

In conclusion, one would hope that there will be available in the near future an accurate, unambiguous measurement of the structure of liquid metals using neutron and X-ray scattering techniques. These results could form the basis of a computation of the interatomic potential using by preference a more sophisticated theory than those available at present. Such a calculation should be accompanied by a thorough investigation of the variation permitted to the interatomic potential while it still reproduces the experimental results to within their estimated precision. The same calculation should be applied to estimations of inert gas liquid-state potentials from structure data as a check on the theory and on the free electron effects. Only then can the pair interaction be concluded to have been unambiguously established within the limits of the theory. With the volume of effort presently dedicated to the study of liquid metals, the picture will no doubt become rapidly clearer. Although the results to date have been rather disappointing, this may yet become a good method of obtaining a pair interaction potential in the near-equilibrium region of interatomic separation.

REFERENCES

1. N. H. March, "Liquid Metals." Pergamon Press, Oxford, 1968.
2. P. A. Egelstaff, "An Introduction to the Liquid State." Academic Press, New York, 1967.
3. *Proc. Int. Conf. Properties Liquid Metals, Brookhaven, 1966*; *Advan. Phys.* **16** (1967).
4. M. Born and H. S. Green, *Proc. Roy. Soc.* **A188**, 10 (1946).
5. J. K. Percus and G. J. Yevick, *Phys. Rev.* **110**, 1 (1958).
6. M. D. Johnson, P. Hutchinson, and N. H. March, *Proc. Roy. Soc.* **A282**, 283 (1964).
7. A. Paskin, *Advan. Phys.* **16**, 223 (1967).
8. B. R. Orton, B. A. Shaw, and G. I. Williams, *Acta Mett.* **8**, 177 (1960).
9. A. Paskin and A. Rahman, *Phys. Rev. Lett.* **16**, 300 (1966).
10. W. S. Gingrich and L. Heaton, *J. Chem. Phys.* **34**, 873 (1961).
11. P. Ascarelli, *Phys. Rev.* **143**, 36 (1966).
12. D. M. North, J. E. Enderby, and P. A. Egelstaff, *J. Phys. C* **1**, 1075 (1968).
13. P. C. Gehlen and J. E. Enderby, *J. Chem. Phys.* **51**, 547 (1969).
14. N. W. Ashcroft and J. Leckner, *Phys. Rev.* **145**, 83 (1966).
15. A. Rahman, *Phys. Rev.* **136**, A405 (1964).
16. D. Schiff, *Phys. Rev.* **186**, 151 (1969), and private communication.
17. R. Pick, *J. Phys.* **28**, 539 (1967).

CHAPTER X

THE APPLICATION OF
INTERATOMIC POTENTIALS

10.1   WHICH POTENTIAL?

One criticism which might be leveled against this monograph so far is that it has not been sufficiently critical or discriminating. It has described a considerable number of potentials without saying a great deal about their merits for various applications. The reason is simply that our information, both experimental and theoretical, is so limited that there is often not much reason to choose one particular potential form in preference to the other forms available. This statement is true for interactions both at equilibrium in the crystal and at higher energies involved in atomic collisions. Furthermore, although applications of pair potentials have been mentioned briefly at various points through the book, no attempt has been made to describe their use in any detail. In fact, not only may the interatomic potential be used to calculate values for some other atomic property, but the results when compared with experiment can cast some light on the potential itself, as for example in the case of computer simulations of liquid metals (Section 9.7). Accordingly, in this final chapter we shall consider some applications of interatomic potentials, attempting to discriminate between the various types of potential available and to suggest which form or forms are physically the most reasonable. Some reviewing or summarizing of what has been said in earlier parts of the book is inevitable, but this is not inappropriate in a concluding chapter.

10.2 WHY A PAIR INTERACTION?

As was explained in Chapter V, the description of an assembly of atoms by means of a sum of pair interactions between them is valid only under quite stringent conditions since the total energy of the system contains other terms due to the interaction of the atomic electrons and indeed due to n-atom interactions which are neglected in the second-order perturbation expansion. In the case of the solid at equilibrium, therefore, care must be taken in any calculation involving pair potentials that their conditions of validity are fulfilled. Whether the results of such a calculation may be trusted depends not only on the accuracy of the chosen potential but on the nature of the solid and on the situation to which the potential is applied. Frequently pair interactions are used simply because a more accurate description of the solid would become rapidly too complicated for practical purposes; this is particularly true of defect calculations in crystals, which we shall be considering in Section 10.5. As was pointed out in the previous chapter, a good description of a liquid metal really would require three-body distribution functions and consequently three-body forces, but in current theories these are expressed in terms of a product of the pair distribution functions.

When we move up to a higher-energy range such as would be involved in atoms colliding in a solid or in a vacuum, the two-body interaction is less questionable, mainly because the repulsive force due to the overlap of closed electron shells increases rapidly with decreasing separation so that the effects of the conduction electron gas or of n-body interactions in the solid, for example, are negligible by comparison. The question remains as to the choice of which potential best represents the interaction at a given energy. This question is further complicated as the energy increases, because the interpenetration of the electron shells becomes such that the repulsion between the nuclei enters the picture.

To an increasing extent application of interatomic potentials to atomic properties requires the use of the computer, whether as a sophisticated calculating machine to solve theoretical equations, or in a complete simulation of a given physical situation in a model assembly of atoms. This gives greater flexibility in the choice of the potential, since purely analytical theory requires generally rather simple and often unrealistic forms of potential in order to carry out the analytical solution. With the computer available, complicated analytical and even numerical forms of potential may be handled with no great difficulty.

We shall consider two principal topics in this chapter, connected with the application of interatomic potentials—atomic collisions and radiation damage studies, which involve the high-energy potential, and crystal defect studies, which require the equilibrium or near-equilibrium part of the potential.

10.3   RADIATION DAMAGE AND ATOMIC COLLISION STUDIES

When a solid material is subjected to incident radiation, the nature and concentration of ensuing damage depend on the characteristics of the radiation and on the type of solid [1, 2]. The mechanisms of damage involve elastic energy transfer processes, electronic excitation and ionization and in some cases nuclear transmutation. In general, irradiation results in the direct or indirect displacement of atoms of the solid from their equilibrium positions. These atoms come to rest through interaction with other atoms of the solid and the initial energy of the primary knock-on atom (PKA) struck by the incident radiation is dissipated through a multiple displacement event known as an atomic displacement cascade. When the average energy of the affected atoms drops to a value comparable with the thermal energy of the undisturbed crystal atoms the final configuration relaxes to a stable damage state which depends on the temperature, background defect concentration, and electronic nature of the solid.

The development of such a multiple atomic displacement event, especially in a crystalline solid, is difficult to treat by analytical methods. Some efforts have been made, ranging from elementary descriptions [3, 4] to rather involved treatments requiring the use of the computer to solve the integral equations obtained [5, 6], but so far none of the theories have been able to take the crystallinity of the material into consideration. They all require a form of atomic interaction—in the case of the more elementary theories a simple hard-sphere potential is used, while for the more involved treatments the integral equations can be solved numerically for a wide range of potentials.

The penetration of charged particles (or indeed uncharged particles such as neutrons) in solids is less complicated to treat analytically, and the use of given collision cross sections permits an estimation of the range and straggling of a given particle in any particular solid. These cross sections generally depend on an assumed form of interatomic potential (see Chapter VII), which should be reasonably valid over the range of collision energies expected. One particular type of penetration is the channeling of charged particles in regions of low atomic density between close-packed rows or planes of a crystal lattice, already mentioned in Chapter IV. Analytical theories have been developed to describe the channeling phenomenon [7], and indeed one such theory, outlined in Appendix 4, demonstrates how to use known experimental stopping cross sections for channeled ions to obtain information regarding the screened potential of crystal atoms.

With the advent of large modern digital computers, a powerful new tool became available for the study of radiation damage in crystals. It is possible to simulate an assembly of atoms in any given crystallographic configuration, the atoms interacting with a chosen interatomic potential. The simulation technique may be applied to various phenomena such as high energy atomic displacement cascades, low energy dynamics, thermal vibrations or static defect properties. In the next section we shall describe the basis of the simulation method, showing how the interatomic pair potential is applied, and then go on in the following three sections to consider several uses of simulation techniques and the potentials which are used in these investigations. This is not intended to be an extensive review of the detailed methods and results obtained; for further information the reader is referred to the original articles and to other reviews of the topic [8, 9].

10.4 ATOMISTIC COMPUTER SIMULATIONS

The interatomic potential is basic to any computer simulation of an array of atoms. Although it has recently been suggested that some direction dependence could be introduced relatively simply [10], most simulations have made use of a simple central pair interaction between the atoms, the nature of which depends on the topic under study.

The simulated array may have any shape and atomic distribution desired, whether solid, liquid or gas, crystalline or amorphous. Having chosen a suitable form of potential, the force between two atoms i and j of the array is given by

$$\mathbf{F}_{ij} = -(\partial V/\partial \mathbf{r}_{ij}) \tag{10.1}$$

and the total force acting on any one atom is obtained by vectorial summation of these pair forces for this atom.

The simulation proceeds through a series of time-steps, each of which consists of the following:

1. summation of the pairwise forces $\mathbf{F}_{ij}$ for each atom, giving $\mathbf{F}_i$;
2. calculation of new velocities $\mathbf{v}_i$ and positions $\mathbf{r}_i$ at the end of the time-step;

3. movement of atoms to their new positions;
4. test of energy conservation.

Of these the most time-consuming stage is the first since it involves a search for neighbors and a potential calculation for a large number of atom pairs. The total number depends on the potential cutoff or distance at which $V(r)$ becomes negligible by comparison with the potential between nearest neighbors. It is evidently desirable to minimize this cutoff distance subject to the condition that accuracy is not lost.

For a finite bounded array, it is necessary to include criteria which simulate the presence of crystal lattice outside the array. These depend on the nature of the problem under consideration. If we are studying the lattice dynamics of atoms of the crystallite at thermal energies the usual method is to use periodic boundary conditions; thus the array is effectively surrounded by identical arrays and its total energy remains constant with time. If on the other hand a radiation damage event initiated in the crystallite is followed by simulation, it would be unrealistic to confine the energy to a small array. The best method of treating the energy dissipation is to apply to the surface atoms a combination of elastic forces which represent their interaction with imaginary external neighbors, and velocity-dependent viscous forces which permit energy loss from the array.

In a dynamical simulation one or more atoms are given initial velocity and the evolution of the system is then followed as a function of time. The atoms are assumed to obey Newtonian mechanics, in which case the equations of motion are given by

$$\dot{\mathbf{v}}_i = (1/m)\mathbf{F}_i = -(1/m)\sum_{i \neq j} \partial V/\partial \mathbf{r}_{ij}, \qquad i = 1, \ldots, n \tag{10.2}$$

where there are n atoms of mass m in the array and $\mathbf{v}_i$ is the velocity of the ith atom. Integration of the n coupled differential equations may be performed using a Taylor expansion over small time intervals Δt:

$$\mathbf{r}_i(t + \Delta t) = \mathbf{r}_i(t) + \mathbf{v}_i(t)\,\Delta t + \tfrac{1}{2}(\mathbf{F}_i/m)(\Delta t)^2 + \cdots \tag{10.3}$$

$$\mathbf{v}_i(t + \Delta t) = \mathbf{v}_i(t) + (\mathbf{F}_i/m)\,\Delta t + \cdots \tag{10.4}$$

The magnitude of Δt depends on the kinetic energies of the moving atoms, and must be such that the force $\mathbf{F}_i$ changes little during the time interval ($\sim 10^{-15}$–10^{-14} sec). The number of terms included in the summation depends on the accuracy required and on the value chosen for Δt. It is possible to reduce the number of terms required by defining the velocity on odd half-integer units of Δt and the position on integer units. Then

$$\mathbf{v}_i(t + \Delta t/2) = \mathbf{v}_i(t - \Delta t/2) + (\mathbf{F}_i(t)/m)\,\Delta t \tag{10.5}$$

$$\mathbf{r}_i(t + \Delta t) = \mathbf{r}_i(t) + \mathbf{v}_i(t + \Delta t/2)\,\Delta t \tag{10.6}$$

A check on the choice of Δt and on the integration procedure should be included in any dynamic simulation, by requiring that the total potential and kinetic energy should be conserved.

Greater accuracy than obtainable by the above techniques is provided by predicter-corrector methods. If we denote by $\mathbf{r}_i^{(p)}(t + \Delta t)$ and $\mathbf{v}_i^{(p)}(t + \Delta t)$ the predicted values based on known values of these quantities at time earlier than $(t + \Delta t)$, and $\mathbf{r}_i^{(c)}(t + \Delta t)$, $\mathbf{v}_i^{(c)}(t + \Delta t)$ the corresponding corrected values, the relevant equations are

$$\mathbf{r}^{(p)}(t + \Delta t) = \mathbf{r}_i(t - \Delta t) + 2 \Delta t\, v_i(t) \tag{10.7}$$

$$\dot{\mathbf{v}}_i^{(p)}(t + \Delta t) = (1/m)\mathbf{F}_i(t + \Delta t) \tag{10.8}$$

$$\mathbf{v}_i^{(c)}(t + \Delta t) = \mathbf{v}_i(t) + (\Delta t/2)\{\dot{\mathbf{v}}_i^{(p)}(t + \Delta t) + \dot{\mathbf{v}}_i(t)\} \tag{10.9}$$

$$\mathbf{r}_i^{(c)}(t + \Delta t) = \mathbf{r}_i(t) + (\Delta t/2)\{\mathbf{v}_i^{(c)}(t + \Delta t) + \mathbf{v}_i(t)\} \tag{10.10}$$

Equations (10.8)–(10.10) constitute an iterative cycle which is repeated until the desired convergence in $\mathbf{r}_i^{(c)}$ and $\mathbf{v}_i^{(c)}$ is obtained.

Computer simulations of atomic ensembles may be approximately divided into four categories:

1. static configurational studies of lattice defects;
2. molecular dynamics of equilibrium and thermal energy processes;
3. dynamics of low energy damage events; and
4. high energy damage and atomic displacement cascades.

Of these the first two are used mainly in the study of lattice defect energies and thermal effects, and liquid properties, while the last two concern radiation damage investigations.

10.5 CRYSTAL DEFECT SIMULATIONS

The purpose of these calculations is generally to find the lattice relaxation and potential energy associated with a defect in a crystal. This may be a point defect such as a vacancy or interstitial, or a line defect such as a dislocation or stacking fault. The crystal is divided into concentric volumes centered on the defect. In the innermost volume the atoms are displaced in such a way as to minimize the potential energy of the system. This is achieved by an iterative process, either moving one atom at a time to its position of minimum energy and cycling this procedure until convergence is attained, or displacing all the atoms simultaneously towards their equilibrium positions each iteration using a matrix inversion procedure. Minimization and relaxation methods have been described by several authors [11–14].

Outside the relaxed volume, the atoms are either held fixed on their lattice sites or slightly displaced according to the elastic constraint imposed by the defect. In some cases the outermost region is simply assumed to be an elastic continuum. Energies of migration of point defects are obtained by relaxing the lattice about the defect in its equilibrium configuration then at the saddle-point for the jump. The difference in potential energy between these two configurations is assumed to be the activation energy required for the jump.

As a variation of the purely static method, a quasi-dynamical approach may also be used to determine the relaxed configuration. First introduced by Gibson *et al.* [15], this involves integrating the equations of motion as described in the previous section for a section of crystal which is "set free" with a defect at the center and all of the atoms initially on their lattice sites. Under normal conditions this would take an excessively large number of time increments Δt to settle down to a stable state, but it can be forced by artificially damping the atom motions. In the computations of Gibson *et al.* such a technique was introduced by setting all atom velocities to zero each time the total energy of the simulated crystallite reached a maximum. A similar method likens the motion of each atom to that of a simple harmonic oscillator, and reduces separately the velocity of the atom to zero each time the total force acting on it becomes perpendicular to its instantaneous velocity. With both these techniques the convergence is considerably more rapid.

There are several difficulties which should be noted regarding the static defect simulation. One problem is inherent in the application of an interatomic potential to the atoms in the immediate neighborhood of a lattice defect. Most pair potentials contain parameters which are adjusted to equilibrium crystal properties, and are therefore valid only in application to a perfect periodic lattice. The details of the assumed potentials could well be altered in the neighborhood of a vacancy for example, and it is very difficult to estimate to what extent this influences the computed results. It is by now well established that the numerical results of static defect simulations are rather potential-sensitive [16, 17].

Assuming, however, that a pair potential is to be used in such a simulation, the question is which type is the most suitable. Evidently since the configuration is never very far from the equilibrium crystal configuration, the potential should be one which applies to this situation. Thus it cannot be purely repulsive, but must possess at least one attractive minimum and possibly some long-range oscillations at greater distances. A considerable amount has been said about such potentials in Chapters IV and VI. The reliability of most of these is questionable, even if they reproduce fairly well some experimental property such as phonon dispersion curves. Possibly the best type to use is that described in Section 4.12, of a composite polynomial nature, where the parameters are adjusted to as many experimental phenomena as possible.

Another problem related to the study of migration energies is the transient nature of the saddle-point, which casts doubt upon the static relaxation about this position. Then there is the role played by thermal vibrations. In reality, the migration of a vacancy is not a simple jump of an atom over a potential barrier, but rather the result of a complex movement of a number of atoms around the defect. For the purposes of diffusion theory, this may be represented by an activation energy, but a static simulation cannot be relied upon to give accurate migration energies even assuming the availability of a good interatomic potential.

10.6 MOLECULAR DYNAMICS

This topic has already been mentioned in Section 9.8 in connection with its application to the structure of liquid metals. To study in detail the effect of thermal vibrations on phenomena related to point-defect migration and diffusion in solids it is necessary to perform a dynamic simulation. The atoms of a volume of crystal (in the form of a cube for cubic crystals) are given a thermal velocity selected at random from a Boltzmann distribution to represent temperature, then set free to interact under the assumed pair potential. Their positions and velocities are followed by numerical integration of their equations of motion through small time increments as described previously. Energy loss is avoided by having periodic boundary conditions which simulate an infinite crystal.

This procedure is evidently rather costly in machine time. If a crystallite contains 1000 atoms (still quite a small volume) the force on each atom due to its interaction with up to 50 neighbors depending on the range of the potential must be evaluated before integrating the equations of motion for each of the 1000 atoms. For any study of vacancy migration the crystal temperature needs to be very high since the jump cannot be "switched on" but must rather await the situation where the vibrations of the atoms surrounding the vacancy are favorably disposed. To obtain any reasonable statistics one might therefore have to follow the system through many thousands of time-steps.

Molecular dynamic simulations were initially applied to liquids, in order to obtain such configurational information as correlation functions, which could be compared with the results of X-ray or neutron scattering data [18]. Recently, crystalline solids have been studied by these techniques. For example, a comparison of the vacancy migration energy in Cu by molecular dynamic and static methods indicates that the latter tends to overestimate the activation energy [19]. This is a relatively new field and will probably develop considerably over the next few years.

The same observations may be made here as in the previous section regarding the application of potentials. It would however be advisable in a molecular dynamic simulation to choose a relatively short-range potential because of the time required for the computational procedure.

10.7   LOW-ENERGY DAMAGE EVENTS

For the range of energy involved in radiation damage (usually greater than 10 eV) dynamic simulations are considerably simpler and more economical. This is partly because we do not have to wait for an event to occur, but can initiate it by giving a selected atom a velocity corresponding to a given energy. In addition, since the energies are so much greater than thermal energies, a threshold may be set and only those atoms receiving energy greater than the threshold considered as moving. Evidently more and more atoms begin to move as the event spreads, but there is nevertheless a large saving in computer time.

The principle of the computation is similar to that described above for molecular dynamic simulations. However the boundary conditions are different, including the velocity-dependent force which simulates the loss of energy from the system. A relatively small number of atoms of the crystallite take part in any damage event, the number depending on the primary energy and the chosen threshold. It should therefore be possible to simplify further the simulation (at least for simple lattices) by storing only the coordinates of the moving atoms as they join the event, rather than those of all of the atoms of the crystallite. This would greatly increase the possible size of the crystallite and hence the possible primary energy. Such a simulation could be used to investigate the validity of displacement cascade models (see next section) at the low energy end of the cascade.

Dynamic simulations have been useful in demonstrating the possibility of focusing and channeling in low-index directions of crystal lattices, which would tend to modify the energy dissipation from that which would be expected in an amorphous solid [15, 20, 21]. Figure 10.1 shows the trajectories of atoms of a crystallite of KCl after one atom has received an energy of 80 eV [21]. The result is a replacement sequence traveling out of the array in the $\langle 110 \rangle$ direction, leaving a vacancy near the position of the primary atom. We note however that this is an event run in a (100) plane, and one intuitively expects directional effects to be more marked with planar geometry than in three dimensions. A small amount of work has been done towards the evaluation of the directional variation of the displacement threshold energy in alpha-iron [20], but the sensitivity of the results to the simulation model and in particular to the interatomic potential was not investigated. This type of computer simulation has lain dormant for several years, although there are signs of a recent renewed interest. A systematic investigation of atomic

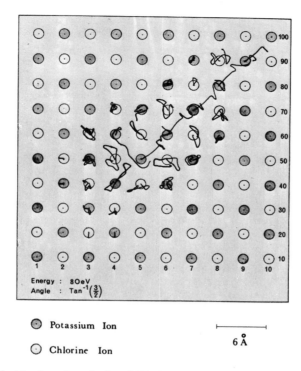

Fig. 10.1. Atomic trajectories in a (100) plane of a KCl crystal, obtained by simulation of the many-body crystal dynamics. (After Torrens and Chadderton [21].)

displacement and focusing in a number of lattices would be desirable, and might help to clarify some controversial points.

If one is interested simply in studying the range of an atom or ion channeled between planes or rows of a lattice, a dynamic simulation becomes again simpler [22, 23]. The atoms struck by the moving particle receive an energy corresponding to the impact parameter and projectile velocity and are subsequently forgotten as the moving atom travels on through the lattice. It is necessary to create new lattice in the path of the channeled particle, the usual method being to displace a rectilinear parallelopiped of lattice by a "leapfrog" technique. Channeling was in fact predicted by computer simulation [24] and subsequently verified by experiment [25], although the phenomenon had been suggested by Stark as long ago as 1912. The computer may be used to find the range distribution of implanted ions or to study the channeling phenomenon per se, although semi-analytical theory for the latter purpose is well developed [7].

The early dynamic simulations used a simple BM repulsive potential, but it soon became obvious that when the energy of the moving atom drops below about 10 eV, the attractive part of the interatomic potential begins to play a role in determining the trajectories. Therefore, it would be best to use an interaction with a minimum, although the part beyond the minimum will have a negligible influence on the results except for energies comparable with those of thermal vibrations in the crystal.

10.8   HIGH-ENERGY ATOMIC DISPLACEMENT CASCADES

Extension of the radiation damage simulation to primary energies in the range involved in for example fast neutron irradiation (up to about 100 keV) may be accomplished at the expense of reducing the atomic collisions to two-body processes. In the previous sections, although a two-body interatomic potential was used, the time-integration of the equations of motion made each collision a many-body process. The two-body collision approximation improves rapidly as the energy increases, since the interatomic potential is a rapidly increasing function and the interaction of a moving atom with the nearest crystal atom will in most cases be large compared to that with the next nearest. However, in some cases where a moving atom passes almost directly between two or more lattice atoms it is necessary to allow for the possibility of nearly simultaneous collisions with these target atoms.

In the simulation of a displacement cascade a primary knock-on atom is given the desired kinetic energy and subsequently interacts with one or more neighbors, displacing them from their lattice sites if they receive sufficient energy. The classical equations for two-body scattering under a central force (see Chapter VII) are solved to find the energy transfer and deflection of the projectile, and the initial direction of motion of the struck atom (see Fig. 7.2). The secondary atoms in their turn strike other atoms and the cascade grows until most of the initial kinetic energy has been dissipated in the lattice. To simulate this event the computer proceeds by selecting for each two-body collision the current most energetic atom in the cascade. A displacement threshold energy (~ 25 eV) is assumed, and if the struck atom receives energy exceeding this threshold it joins the cascade. The simulation continues until all the cascade atoms have kinetic energy less than a specified value. Then some form of close-pair or thermal annealing may be applied to the interstitial-vacancy concentration remaining at the end.

Inelastic energy loss, significant in the kilovolt energy range, can be included by calculating the energy given to electron excitation during a collision and subtracting it from the projectile energy. The effect of finite temperature can be simulated in an adiabatic approximation, where the target atom is given a small thermal displacement and considered stationary for the duration of the collision. A simple system of creating a small section of crystal for each target search avoids having to store the coordinates of a large number of atoms in the computer. Thus there is practically no limit to cascade dimensions in space, and different crystal lattice forms may be easily manipulated. The only effective limitation is the number of atoms taking part in the cascade, and this is quite adequate for primary energies of interest to most radiation damage studies.

Most early work in this topic was undertaken by Beeler and co-workers, who simulated cascades of primary energy up to 30 keV in lattices of BeO, alpha-iron, Cu, and W [26, 27]. No inelastic energy loss was included in their model. A new program for simulating cascades of primary energy up to about 1 MeV in a wide variety of materials including inelastic loss and finite temperature effects has been developed by Torrens and Robinson [28]. An example of a 75-keV cascade in fcc Cu is shown in Fig. 10.2. It shows the

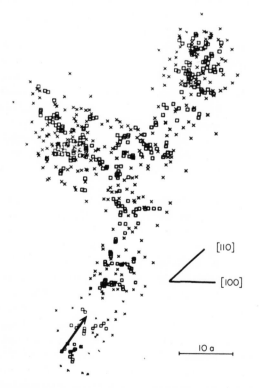

Fig. 10.2. Projection on a (100) plane of an atomic displacement cascade in Cu originating from a 75-keV primary atom (arrowed). Vacancies are depicted by squares and interstitials by crosses. The scale is in units of the lattice parameter a. Crystallographic directions [100] and [110] are indicated (After Torrens and Robinson [28].)

division into two or more branches which is typical of cascades at higher energy.

The main problem of high energy cascade simulation is the low energy end of the cascade below about 100 eV, where many-body effects become significant. It will be necessary to study the model in more detail in this region, perhaps improving it using information gained from dynamic many-body simulations at these relatively low energies.

Since the energies involved in the displacement cascade simulation are above the displacement threshold energy, there is no need to include an attractive part in the interatomic potential. Doing so would only serve to complicate the classical two-body scattering calculation. The choice of the pair interaction must depend on the energies involved in atomic collisions in the cascade, the maximum possible being of course the PKA kinetic energy. As a general rule, one may use a simple BM interaction if the PKA energy is less than or comparable to the value of the BM preexponential constant. For higher-energy cascades, it is necessary to include some form of screened nuclear interaction. This may be accomplished either by joining a screened Coulomb and BM potential at some intermediate point, or by devising a mixed analytical potential which incorporates both types; these potentials were described in Section 4.10. In particular, the Molière approximation to the TF potential (Section 4.11) provides a good approximation at both ends of the energy range, with a suitable choice of screening radius, and is eminently suitable for computational purposes.

10.9   INTERPRETATION OF COMPUTER SIMULATION RESULTS

Computer development during the past decade has been such as to make atomistic computer simulations of many different problems quite feasible. These techniques have in the past frequently been referred to as "computer experiments." The difference between these and real experiments is, of course, that in the former case the laws of nature are provided as input data to as good an approximation as possible, based on current knowledge and machine limitations. It cannot be sufficiently stressed that the results of computer simulations, like those of any other theoretical study, are only as good as the model. Consequently it is essential to study the sensitivity of the results to various aspects of the model. One major uncertainty of this model is the form of the interatomic potential, and very little significance may be attached to the results of any computer simulation which does not look into their sensitivity to the chosen potential. In general experience, the most interesting results from computer simulation are not absolute numerical values for a given quantity, but rather the comparative values of two or more quantities—for example, in the dynamic simulations the directional variation of the atomic displacement threshold energy is of more interest than an absolute value in a given direction, since it tends to be a little less potential-sensitive. Provided that a certain amount of caution is exercised in interpreting the results of computer simulations in terms of reality, there is no reason why these techniques should not be used with considerable success to improve our understanding of problems involving atomic motions in crystals.

10.10 INTERATOMIC POTENTIALS: THE STATE OF THE ART

The purpose of this book has been to discuss the problem of the interaction between atoms over the complete energy range. From a theoretical point of view, the state of the art is satisfactory only at very high interaction energies and very small separations, where the atomic electrons represent a minor perturbation on the strong nuclear repulsion, and at very large distances in vacuo, where the forces between the atoms are van der Waals in nature. In the intermediate energy range where the overlap of closed electron shells is substantial the many-body problem becomes acute and no truly satisfactory method of calculating the potential in this region is presently evident. It is therefore necessary to extrapolate from high and low energy interactions in order to obtain information in this range, which is important for the study of such phenomena as radiation damage in solids. In the low energy region, empirical potentials for solids may be deduced from crystal equilibrium data by parameter adjustment of an assumed form, but it is frequently difficult to have a great deal of confidence in the results. The method of pseudopotentials has introduced a more theoretical approach to the equilibrium potential for restricted classes of metallic crystal, but again the uncertainties are significant. Experimental methods of determining directly the pair potential are few, and those that exist are rather difficult to perform.

On a more optimistic note, the availability of high-speed computers has enabled more experimental information to be manipulated in the creation of more phenomenological potentials, especially in the equilibrium region. Until theoretical understanding makes some considerable progress, this may be the most promising approach. The intermediate energy range still poses some problems, but extrapolation techniques give us a reasonable order-of-magnitude knowledge of the potential.

In view of the uncertainty in the interatomic potential, it is crucial in any theoretical study using a given potential form to investigate the sensitivity of the results obtained to variation of the chosen interaction. This has perhaps not been sufficiently stressed in the past, and frequently results have been published based on a particular interaction without any systematic study of how sensitive these results may be to the model. Even when an observation is made to the effect that the results may be potential-sensitive, they are commonly assumed by later researchers to represent reality. Thus it would be advisable, when a theory requires a two-body interaction between atoms, to perform the calculation using several different potentials or to vary systematically the parameters of the assumed form within a reasonable range. Comparison of the calculated and experimental results may then yield some information on the potential.

REFERENCES

1. L. T. Chadderton, " Radiation Damage in Crystals." Methuen, London, 1965.
2. M. W. Thompson, " Defects and Radiation Damage in Metals." Cambridge Univ. Press, London and New York, 1969.
3. G. H. Kinchin and R. C. Pease, *Rep. Progr. Phys.* **18**, 1 (1955).
4. W. S. Snyder and J. Neufeld, *Phys. Rev.* **97**, 1636 (1955).
5. K. B. Winterbon, P. Sigmund, and J. B. Sanders, *Kgl. Dansk. Vidensk. Selsk. Mat.-Fys. Medd.* **37**, No. 14 (1970).
6. M. T. Robinson, *Phil. Mag.* **12**, 741 (1965); **17**, 639 (1968).
7. J. Lindhard, *Kgl. Dansk. Vidensk. Selsk. Mat.-Fys. Medd.* **34**, No. 14 (1965).
8. J. R. Beeler, Jr., " Physics of Many-Particle Systems " (E. Meeron, ed.), p. 1. Gordon and Breach, New York, 1967.
9. D. Van Vliet, *in* " Studies in Radiation Effects in Solids " (G. J. Dienes and L. T. Chadderton, eds.), Vol. 5. Gordon & Breach, New York (in press).
10. R. A. Johnson, *in* "Interatomic Potentials and Simulation of Lattice Defects" (P. C. Gehlen, J. R. Beeler, Jr., and R. I. Jaffee, eds.), pp. 301–319. Plenum, New York, 1972.
11. R. A. Johnson, *Phys. Rev.* **134**, A1329 (1964).
12. I. M. Torrens and M. Gerl, *Phys. Rev.* **187**, 912 (1969).
13. J. E. Sinclair and H. F. Pollard, *Phys. Lett.* **32A**, 93 (1970).
14. M. J. Norgett, *J. Phys. C* **3**, L1 90 (1970).
15. J. B. Gibson, A. N. Goland, M. Milgram, and G. H. Vineyard, *Phys. Rev.* **120**, 1229 (1960).
16. R. A. Johnson, *Radiat. Effects* **2**, 1 (1969).
17. A. DePino, Jr., D. G. Doran, and J. R. Beeler, Jr., *Radiat. Effects* **3**, 23 (1970).
18. A. Rahman, *Phys. Rev.* **136**, A405 (1964).
19. D. H. Tsai, R. Bullough, and R. C. Perrin, *J. Phys. C* **3**, 2022 (1970).
20. C. Erginsoy, G. H. Vineyard, and A. Englert, *Phys. Rev.* **133**, A595 (1964).
21. I. M. Torrens and L. T. Chadderton, *Phys. Rev.* **159**, 671 (1967).
22. I. M. Torrens and L. T. Chadderton, *Canad. J. Phys.* **46**, 1303 (1968).
23. D. V. Morgan and D. Van Vliet, *Canad. J. Phys.* **46**, 503 (1968).
24. M. T. Robinson and O. S. Oen, *Phys. Rev.* **132**, 2385 (1963).
25. G. R. Peircy, M. McCargo, F. Brown, and J. A. Davies, *Canad. J. Phys.* **42**, 1116 (1964).
26. J. R. Beeler, Jr. and D. G. Besco, *J. Appl. Phys.* **34**, 2873 (1963).
27. J. R. Beeler, Jr., *Phys. Rev.* **134**, A530 (1964); **150**, 470 (1966).
28. I. M. Torrens and M. T. Robinson, *in* "Interatomic Potentials and Simulation of Lattice Defects " (P. C. Gehlen, J. R. Beeler, Jr., and R. I. Jaffee, eds.), pp. 423–438. Plenum, New York, 1972.

THE THOMAS–FERMI SCREENING FUNCTION

TABLE OF VALUES OF THE TF SCREENING FUNCTION $\chi(x)$ AND ITS DERIVATIVE[a]

x	$\chi(x)$	$-\chi'(x)$	x	$\chi(x)$	$-\chi'(x)$
0.0000	.100 000 00E + 01	.158 807 10E + 01	.1156	.866 477 61E − 00	.956 497 83E − 00
.0001	.999 842 53E − 00	.156 807 26E + 01	.1296	.853 312 90E − 00	.924 600 76E − 00
.0004	.999 375 44E − 00	.154 808 36E + 01	.1444	.839 862 38E − 00	.893 441 41E − 00
.0009	.998 606 72E − 00	.152 811 31E + 01	.1600	.826 164 92E − 00	.863 028 59E − 00
.0016	.997 544 36E − 00	.150 817 01E + 01	.1764	.812 257 47E − 00	.833 368 64E − 00
.0025	.996 196 30E − 00	.148 826 32E + 01	.1936	.798 175 11E − 00	.804 465 59E − 00
.0036	.994 570 47E − 00	.146 840 08E + 01	.2116	.783 951 04E − 00	.776 321 34E − 00
.0049	.992 674 76E − 00	.144 859 12E + 01	.2304	.769 616 64E − 00	.748 935 79E − 00
.0064	.990 517 02E − 00	.142 884 21E + 01	.2500	.755 201 47E − 00	.722 306 98E − 00
.0081	.988 105 06E − 00	.140 916 13E + 01	.2704	.740 733 34E − 00	.696 431 29E − 00
.0100	.985 446 61E − 00	.138 955 61E + 01	.2916	.726 238 34E − 00	.671 303 52E − 00
.0121	.982 549 38E − 00	.137 003 37E + 01	.3136	.711 740 89E − 00	.646 917 07E − 00
.0144	.979 421 00E − 00	.135 060 08E + 01	.3364	.697 263 79E − 00	.623 264 02E − 00
.0169	.976 069 01E − 00	.133 126 42E + 01	.3600	.682 828 26E − 00	.600 335 35E − 00
.0196	.972 500 90E − 00	.131 203 01E + 01	.4096	.654 159 − 31E 00	.556 609 81E − 00
.0225	.968 724 10E − 00	.129 290 47E + 01	.4624	.625 874 54E − 00	.515 649 26E − 00
.0256	.964 745 90E − 00	.127 389 38E + 01	.5184	.598 092 91E − 00	.477 351 06E − 00
.0289	.960 573 57E − 00	.125 500 30E + 01	.5776	.570 913 67E − 00	.441 604 05E − 00
.0324	.956 214 25E − 00	.123 623 77E + 01	.6400	.544 418 17E − 00	.408 291 20E − 00
.0361	.951 675 00E − 00	.121 760 29E + 01	.7056	.518 671 63E − 00	.377 291 86E − 00
.0400	.946 962 77E − 00	.119 910 35E + 01	.7744	.493 724 84E − 00	.348 483 68E − 00
.0441	.942 084 44E − 00	.118 074 40E + 01	.8464	.469 615 81E − 00	.321 744 21E − 00
.0484	.937 046 76E − 00	.116 252 90E + 01	.9216	.446 371 26E − 00	.296 952 28E − 00
.0529	.931 856 39E − 00	.114 446 25E + 01	1.0000	.424 008 05E − 00	.273 989 05E − 00
.0576	.926 519 87E − 00	.112 654 84E + 01	1.0500	.410 645 48E − 00	.260 674 56E − 00
.0625	.921 043 65E − 00	.110 879 05E + 01	1.1000	.397 925 30E − 00	.248 278 12E − 00
.0676	.915 434 06E − 00	.109 119 21E + 01	1.1500	.385 803 79E − 00	.236 714 47E − 00
.0729	.909 697 31E − 00	.107 375 66E + 01	1.2000	.374 241 23E − 00	.225 908 59E − 00
.0784	.903 839 50E − 00	.105 648 70E + 01	1.2500	.363 201 41E − 00	.215 794 13E − 00
.0841	.897 866 63E − 00	.103 938 62E + 01	1.3000	.352 651 28E − 00	.206 312 18E − 00
.0900	.891 784 56E − 00	.102 245 67E + 01	1.3500	.342 560 53E − 00	.197 410 26E − 00
.1024	.879 315 73E − 00	.989 121 20E − 00	1.4000	.332 901 37E − 00	.189 041 43E − 00

x		
1.5000	.314 777 46E − 00	.173 738 80E − 00
1.6000	.298 097 71E − 00	.160 115 01E − 00
1.7000	.282 706 44E − 00	.147 933 39E − 00
1.8000	.268 469 51F − 00	.136 998 44E − 00
1.9000	.255 270 65E − 00	.127 147 29E − 00
2.0000	.243 008 51E − 00	.118 243 19E − 00
2.1000	.231 594 32E − 00	.110 170 54E − 00
2.2000	.220 949 98E − 00	.102 830 98E − 00
2.3000	.211 006 50E − 00	.961 403 50E − 01
2.4000	.201 702 70E − 00	.900 262 76E − 01
2.5000	.192 984 12E − 00	.844 261 87E − 01
2.6000	.184 802 15E − 00	.792 857 63E − 01
2.7000	.177 113 23E − 00	.745 576 47E − 01
2.8000	.169 878 26E − 00	.702 003 88E − 01
2.9000	.163 062 01E − 00	.661 775 80E − 01
3.0000	.156 632 67E − 00	.624 571 31E − 01
3.2000	.144 822 25E − 00	.558 130 27E − 01
3.4000	.134 247 00E − 00	.500 771 16E − 01
3.6000	.124 741 04E − 00	.450 976 30E − 01
3.8000	.116 165 70E − 00	.407 527 38E − 01
4.0000	.108 404 26E − 00	.369 437 58E − 01
4.2000	.101 357 87E − 00	.335 900 97E − 01
4.4000	.949 423 09E − 01	.306 254 44E − 01
4.6000	.890 854 40E − 01	.279 948 61E − 01
4.8000	.837 251 63E − 01	.256 525 43E − 01
5.0000	.788 077 79E − 01	.235 600 75E − 01
5.2000	.742 864 47E − 01	.216 850 62E − 01
5.4000	.701 210 97E − 01	.200 000 50E − 01
5.6000	.662 755 27E − 01	.184 816 57E − 01
5.8000	.627 186 65E − 01	.171 098 84E − 01
6.0000	.594 229 49E − 01	.158 675 50E − 01
6.5000	.521 729 37E − 01	.132 356 07E − 01
7.0000	.460 978 19E − 01	.111 425 32E − 01

x		
7.5000	.409 624 66E − 01	.945 826 46E − 02
8.0000	.365 872 55E − 01	.808 860 30E − 02
8.5000	.328 330 89E − 01	.696 416 38E − 02
9.0000	.295 909 35E − 01	.603 307 47E − 02
9.5000	.267 743 90E − 01	.525 603 00E − 02
10.0000	.243 142 93E − 01	.460 288 19E − 02
11.0000	.202 503 65E − 01	.357 981 52E − 02
12.0000	.170 639 22E − 01	.283 053 64E − 02
13.0000	.145 265 18E − 01	.227 052 46E − 02
14.0000	.124 784 06E − 01	.184 450 14E − 02
15.0000	.108 053 59E − 01	.151 532 31E − 02
16.0000	.942 407 89E − 02	.125 743 53 E − 02
17.0000	.827 276 39E − 02	.105 288 68E − 02
18.0000	.730 484 59E − 02	.888 831 11E − 03
19.0000	.648 474 64E − 02	.755 921 42E − 03
20.0000	.578 494 12E − 02	.647 254 33E − 03
21.0000	.518 389 34E − 02	.557 661 58E − 03
22.0000	.466 457 58E − 02	.483 225 74E − 03
23.0000	.421 339 81E − 02	.420 943 70E − 03
24.0000	.381 941 81E − 02	.368 489 22E − 03
25.0000	.347 375 44E − 02	.324 043 00E − 03
26.0000	.316 914 44E − 02	.286 169 52E − 03
27.0000	.289 960 77E − 02	.253 726 72E − 03
28.0000	.266 018 79E − 02	.225 799 01E − 03
29.0000	.244 675 26E − 02	.201 647 09E − 03
30.0000	.225 583 66E − 02	.180 670 01E − 03
32.0000	.193 032 55E − 02	.146 361 06E − 03
34.0000	.166 519 08E − 02	.119 884 59E − 03
36.0000	.144 695 44E − 02	.991 771 75E − 04
38.0000	.126 561 39E − 02	.827 855 36E − 04
40.0000	.111 363 56E − 02	.696 680 29E − 04
42.0000	.985 269 00E − 03	.590 661 39E − 04
44.0000	.876 070 64E − 03	.504 195 38E − 04

[a] P. J. Rijnierse, in "Studies in Radiation Effects in Solids" (G. J. Dienes and L. T. Chadderton, eds.), Vol. 5. Gordon & Breach, New York (in press).

Appendix 1 (*continued*)

x	$\chi(x)$	$-\chi'(x)$	x	$\chi(x)$	$-\chi'(x)$
46.0000	.782 569 14E − 03	.433 088 48E − 04	230.0000	.974 309 01E − 05	.120 875 22E − 06
48.0000	.702 024 72E − 03	.374 164 00E − 04	240.0000	.862 806 70E − 05	.102 744 31E − 06
50.0000	.632 254 78E − 03	.324 989 02E − 04	250.0000	.767 729 08E − 05	.878 946 80E − 07
52.0000	.571 505 54E − 03	.283 681 26E − 04	260.0000	.686 154 84E − 05	.756 379 15E − 07
54.0000	.518 356 49E − 03	.248 770 65E − 04	270.0000	.615 765 44E − 05	.654 485 28E − 07
56.0000	.471 648 54E − 03	.219 099 04E − 04	280.0000	.554 704 71E − 05	.569 213 13E − 07
58.0000	.430 429 33E − 03	.193 746 55E − 04	290.0000	.501 474 64E − 05	.497 408 61E − 07
60.0000	.393 911 37E − 03	.171 977 00E − 04	300.0000	.454 857 20E − 05	.436 594 96E − 07
65.0000	.319 143 29E − 03	.129 604 11E − 04	320.0000	.377 646 38E − 05	.340 493 89E − 07
70.0000	.262 265 30E − 03	.995 653 34E − 05	340.0000	.316 996 37E − 05	.269 470 14E − 07
75.0000	.218 210 43E − 03	.777 797 47E − 05	360.0000	.268 689 55E − 05	.216 058 08E − 07
80.0000	.183 545 76E − 03	.616 619 55E − 05	380.0000	.229 735 43E − 05	.175 262 91E − 07
85.0000	.155 887 83E − 03	.495 261 18E − 05	400.0000	.197 973 26E − 05	.143 668 23E − 07
90.0000	.133 545 83E − 03	.402 447 37E − 05	420.0000	.171 815 18E − 05	.118 890 56E − 07
95.0000	.115 297 15E − 03	.330 466 01E − 05	440.0000	.150 076 45E − 05	.992 371 54E − 08
100.0000	.100 242 57E − 03	.273 935 11E − 05	460.0000	.131 860 86E − 05	.834 862 85E − 08
105.0000	.877 104 51E − 04	.229 030 47E − 05	480.0000	.116 481 92E − 05	.707 431 70E − 08
110.0000	.771 921 84E − 04	.192 990 23E − 05	500.0000	.103 407 72E − 05	.603 436 34E − 08
115.0000	.682 977 13E − 04	.163 790 00E − 05	520.0000	.922 217 65E − 06	.517 885 89E − 08
120.0000	.607 244 54E − 04	.139 925 83E − 05	540.0000	.825 947 95E − 06	.446 987 33E − 08
125.0000	.542 351 97E − 04	.120 266 54E − 05	560.0000	.742 641 55E − 06	.387 827 77E − 08
130.0000	.486 421 71E − 04	.103 951 57E − 05	580.0000	.670 185 90E − 06	.338 148 50E − 08
140.0000	.395 741 39E − 04	.788 564 47E − 06	600.0000	.606 868 77E − 06	.296 182 25E − 08
150.0000	.326 339 64E − 04	.609 139 95E − 06	650.0000	.479 984 46E − 06	.216 546 95E − 08
160.0000	.272 310 37E − 04	.478 074 16E − 06	700.0000	.386 176 52E − 06	.161 983 22E − 08
170.0000	.229 613 51E − 04	.380 511 34E − 06	750.0000	.315 325 80E − 06	.123 583 41E − 08
180.0000	.195 421 02E − 04	.306 664 32E − 06	800.0000	.260 813 73E − 06	.959 243 86E − 09
190.0000	.167 712 48E − 04	.249 929 01E − 06	850.0000	.218 186 65E − 06	.755 927 24E − 09
200.0000	.145 018 03E − 04	.205 753 23E − 06	900.0000	.184 372 42E − 06	.603 766 18E − 09
210.0000	.126 250 79E − 04	.170 938 68E − 06	950.0000	.157 205 04E − 06	.488 057 17E − 09
220.0000	.110 599 15E − 04	.143 199 26E − 06	1000.0000	.135 127 48E − 06	.398 801 07E − 09

226

HARTREE DIELECTRIC SCREENING
OF THE PSEUDOPOTENTIAL

The conduction electrons in a metal interact not only with the ion cores but with each other through Coulomb potentials. We can estimate the effect of the electron interactions, treated as a perturbation, in a self-consistent field calculation of the Hartree type, which neglects exchange and correlation. The presence of the other electrons will add an extra term W_e to the total pseudopotential as seen by a conduction electron:

$$W_s(r) = W_b(r) + W_e(r) \tag{A2.1}$$

To find the value of W_e we must compute the electron density $\rho(r)$ self-consistently by a first-order perturbation calculation. The result of this calculation is

$$\rho(r) = \sum_{q \neq 0} \rho(q)e^{i\mathbf{q}\cdot\mathbf{r}} \tag{A2.2}$$

where

$$\rho(q) = \frac{1}{2\pi^3} \int \frac{\langle \mathbf{k} + \mathbf{q} | W | \mathbf{k} \rangle}{(\hbar^2/2m)(k^2 - |\mathbf{k} + \mathbf{q}|^2)} \, d^3\mathbf{k} \qquad (A2.3)$$

For a local pseudopotential this reduces to

$$\rho(q) = \frac{mW(q)}{\pi^3\hbar^2} \int \frac{d^3\mathbf{k}}{k^2 - |\mathbf{k} + \mathbf{q}|^2} \qquad (A2.4)$$

The vector $\mathbf{k}$ is then resolved into components $\mathbf{k}_\parallel$ and $\mathbf{k}_\perp$ parallel and perpendicular to $\mathbf{q}$ and the integration carried out over the Fermi sphere:

$$\rho(q) = -\frac{mW(q)}{\pi^2\hbar^2} \int_{-k_F}^{k_F} \frac{k_F^2 - k_\parallel^2}{q^2 + 2qk_\parallel} \, dk_\parallel \qquad (A2.5)$$

This gives upon integrating

$$\rho(q) = -\frac{mk_F W(q)}{2\pi^2\hbar^2} \left\{ 1 + \frac{4k_F^2 - q^2}{2k_F q} \ln \left| \frac{2k_F + q}{2k_F - q} \right| \right\} \qquad (A2.6)$$

Self-consistency is established by means of Poisson's equation:

$$W_e(q) = (4\pi e^2/q^2)\rho(q) \qquad (A2.7)$$

Then substituting into Eq. (A2.1) we have finally

$$W_s(q) = W_b(q)/\varepsilon(q) \qquad (A2.8)$$

where

$$\varepsilon(q) = 1 + \frac{2k_F m e^2}{\pi\hbar^2 q^2} \left\{ 1 + \frac{4k_F^2 - q^2}{4k_F q} \ln \left| \frac{2k_F + q}{2k_F - q} \right| \right\} \qquad (A2.9)$$

Because the calculation involves only a first-order perturbation this is a linear screening function. In this approximation, each ion may also be screened independently using the same function, so that

$$w_s(q) = w_b(q)/\varepsilon(q) \qquad (A2.10)$$

The nature of the dielectric function $\varepsilon(q)$ and its effect on the pseudopotential is discussed in the text (see Section 5.6).

THE COHESION OF IONIC CRYSTALS

In Chapter I we discussed briefly the different types of binding between atoms of a solid, the simplest of which was the ionic bond. Such crystals are made up of positive and negative ions held together largely by Coulomb forces. It is perhaps correct to say that we have more accurate information on the interionic forces in this type of solid than on those for any other type, and that the two-body approximation has greatest validity in ionic crystals.

The interionic potential in an ionic crystal may be expressed as the sum of four different terms—the electrostatic, closed shell repulsive, van der Waals, and zero-point energy. Thus the pair potential between ions i and j in a crystal may be written

$$V(r_{ij}) = \pm e^2/r_{ij} + a/r_{ij}^n - b/r_{ij}^6 + E_v \qquad (A3.1)$$

The last term is small compared to the other three.

The repulsive term has been written in the power-law form of the Born model [1], but can also be expressed as a BM exponential repulsion (see Chapter IV):

$$V(r_{ij})_{\text{repulsive}} = a \exp\{(r_i + r_j - r)/\rho\} \tag{A3.2}$$

Here r_i and r_j are the ionic radii of the interacting ions. A good recent set of values of the ionic radii is given in a review by Tosi [2]. This repulsive force is due to the overlap of closed shells of the neighboring ions and decays very rapidly. In the Born model the constants a and n or ρ are found from crystal data (n is normally about 10). The repulsive potential gives only a minor contribution to the total energy of the crystal and may be neglected for other than nearest-neighbor interactions.

The electrostatic potential energy, which stems from the first term in Eq. (A3.1), is quite straightforward, being a simple Coulomb interaction. When the pair terms are summed over the crystal to find the electrostatic contribution to the cohesive energy, the result is usually written in the form of a Madelung summation:

$$E_{\text{el.}} = -\alpha e^2/r_0 \tag{A3.3}$$

where r_0 is the nearest-neighbor distance. The constant α depends only on crystal structure and is the result of summation of the electrostatic interactions over groups of lattice ions chosen to ensure convergence. In the case of the simple cubic (or NaCl type) alkali halides, α is 1.74756.

The van der Waals term gives only a weak contribution to the pair potential and to the cohesive energy. There is fairly good agreement with experimental values of the lattice energy even if this term is not included in the theoretical calculation. The constant b may be calculated theoretically from the polarizabilities and excitation energies of the atoms.

When considering the dynamical properties of ionic crystal lattices it is necessary to take into account the polarization induced by the displacement deformation of an ion. The effect of the deformation dipole induced by ionic displacement is to reduce somewhat the *effective* charge of the ion in its interaction with another ion (though not when we are estimating the total static lattice cohesive energy). This effect has been calculated by Szigeti [3] who gives the calculated effective charges for a number of different ionic solids.

This is only intended to be a very brief outline of the interionic potentials for ionic crystals. These solids are well documented in almost any solid state textbook and good reviews are widely available [2, 4–7].

REFERENCES

1. M. Born, "Atomtheorie des festen Zustandes." Teubner, Leipzig, 1923.
2. M. P. Tosi, *Solid State Phys.* **16**, 1 (1964).
3. B. Szigeti, *Proc. Roy. Soc.* **A204**, 51 (1950).

4. M. Born and K. Huang, " Dynamical Theory of Crystal Lattices." Oxford Univ. Press, London and New York, 1954.
5. N. F. Mott and R. W. Gurney, "Electronic Processes in Ionic Crystals," Chapter 1. Oxford Univ. Press, London and New York, 1948.
6. C. Kittel, " Introduction to Solid State Physics," Chapter 3. Wiley, New York, 1956.
7. L. W. Barr and A. B. Lidiard, *in* " Physical Chemistry " (W. Jost, ed.),Vol. 10, Chapter 3. Academic Press, New York, 1970.

INTERATOMIC POTENTIALS DERIVED FROM PLANAR CHANNELING DATA

The phenomenon of channeling was mentioned briefly in Chapter IV, and one method was described by which some rather indirect information about the Z-dependence of the interatomic potential could be obtained. Some recent experimental work of a much more precise nature has made possible an interpretation which links fairly decisively the interaction potential with the channeled ion energy loss in a thin monocrystalline film. Although still in its early stages of application, this method has had sufficient success to indicate that it will be very useful in probing the potential [1, 2].

Experiments have been performed for He, O, and iodine ions with initial energy ranging from 3 to 60 MeV incident on a single crystal of Au [3–5], and for 400-keV protons on Si [6]. The accelerated ions enter the crystal at an angle such that they are constrained to move between close-packed crystal planes, undergoing only grazing collisions with atoms of these planes before leaving the back surface of the crystal and entering the detector. The most

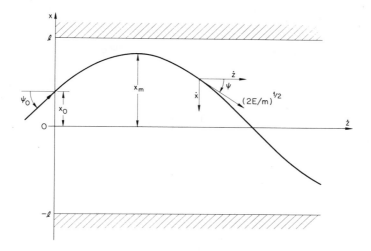

Fig. A4.1. Schematic representation of planar channeling, showing the coordinates used in the analysis. (After Robinson [2].)

important aspect of the experimental work is the high degree of precision in beam collimation and detection. The ion beam had a spread of 0.05° and the detector had an acceptance angle of $\sim 0.01°$ and an energy analyzer with a resolution $\lesssim 0.1$ MeV. This permitted a determination of the oscillation frequency and stopping-power of the ions in the planar channel, and experimental data revealed a direct proportionality between these quantities. Thus, assuming some form of interatomic potential, the transverse oscillations could be analyzed to yield a theoretical estimate of the stopping-power, which could then be compared with experiment to check the potential.

The theory connecting the potential and channeled ion energy loss has been given by Robinson [2]. When an ion is traveling in a channel parallel to crystal planes and making only glancing collisions, the individual ion–atom potential $V(r)$ may be replaced by a planar continuum potential:

$$V_1(\bar{x}) = 4\pi\kappa\rho l \int_{\bar{x}}^{\infty} rV(r)\, dr \qquad (A4.1)$$

Here ρ is the atomic density of the target, l the half-width of the channel, $\bar{x}$ the distance from the ion to the atomic plane, and κ is a factor allowing for the possibility that the atomic density of the plane differs from $2\rho l$. When a pair of planes is present (see Fig. A4.1) the net potential acting on the ion is

$$V_2(x) = V_1(l + x) + V_1(l - x), \qquad -l \le x \le \qquad (A4.2)$$

If $S(x, E)$ is the stopping power of the medium and ψ is the angle between the instantaneous trajectory of the ion and the plane, the equations of motion of an ion of mass m are

$$m\ddot{x} + (d/dx)\{V(x)\} + S(x, E) \sin \psi = 0 \qquad (A4.3)$$

$$m\ddot{z} + S(x, E) \cos \psi = 0 \qquad (A4.4)$$

Since ψ is always small for a channeled ion we can neglect the term in $\sin \psi$ so that the stopping-power effectively influences only the longitudinal component. Then from Eq. (A4.3)

$$\dot{x} = (2/m)^{1/2}\{V_2(x_m) - V_2(x)\}^{1/2} \qquad (A4.5)$$

This is an oscillatory motion of amplitude x_m and period τ, given by

$$\tau = (8m)^{1/2} \int_0^{x_m} \{V_2(x_m) - V_2(x)\}^{-1/2} \, dx \qquad (A4.6)$$

Suppose a "frequency" ω is defined by the equation

$$\omega^{-1} = 2 \int_0^{x_m} \{V_2(x_m) - V_2(x)\}^{-1/2} \, dx \qquad (A4.7)$$

If the stopping-power is now written in the form

$$S(x, E) = s_0 + s_1[\sigma(x) - 1] \qquad (A4.8)$$

where $\sigma(0) = 1$, then the mean stopping-power over an integral number of half wavelengths is obtained by averaging:

$$\langle S(x, E) \rangle = \frac{\int_0^{x_m} \{s_0 + s_1[\sigma(x) - 1]\}(dx/\dot{x})}{\int_0^{x_m} dx/\dot{x}} \qquad (A4.9)$$

$$= \alpha + \beta\omega \qquad (A4.10)$$

where

$$\alpha = s_0 - s_1 \qquad (A4.11)$$

and

$$\beta = 2s_1 \int_0^{x_m} \sigma(x)\{V_2(x_m) - V_2(x)\}^{1/2} \, dx \qquad (A4.12)$$

Since it was found experimentally that the stopping-power was proportional to the oscillation frequency, the data provide a measure of α and β.

There are now two methods of finding the potential $V_2(x)$. We may assume a given form for the ion–atom potential $V(r)$, integrate Eq. (A4.1) to find V_1 and V_2 and try to find a function $\sigma(x)$ for which β is constant. Alternatively and more directly, the solution of Eq. (A4.12), assuming β constant, is

$$\sigma(x) = \frac{\beta}{\pi s_1} \frac{d}{dx} \{V_2(x) - V_2(0)\}^{1/2}, \qquad 0 \le x \le l \qquad (A4.13)$$

Then since $V_2(x)$ is an even function:

$$\sigma(0) = \frac{\beta}{2^{1/2}\pi s_1} \{V_2''(0)\}^{1/2} \qquad (A4.14)$$

This may be expressed in terms of a *curvature parameter y*, which measures the curvature of the potential at the center of the channel:

$$y = V_2''(0)/l \qquad (A4.15)$$

In terms of the experimental energy loss spectra, this may be expressed, using Eq. (A4.11) and (A4.14):

$$y = 2\pi^2(s_0 - \alpha)^2/\beta^2 l \qquad (A4.16)$$

Also, from Eqs. (A4.1) and (A4.2),

$$y = 2V_1(l)''/l = -8\pi\kappa\rho\{V(l) + lV'(l)\} \qquad (A4.17)$$

Thus if the channeled ion stopping-power is determined experimentally for several channels of different l, the parameters of an assumed potential can be found.

An independent check of the potential function evaluated in this way comes from the random stopping-power $\hat{S}$ of the nonchanneled ions, given in terms of the measurable quantities by averaging Eq. (A4.8):

$$\hat{S} = s_0 - s_1 + (s_1/l) \int_0^l \sigma(x) \, dx \qquad (A4.18)$$

$$= \alpha + (\beta/\pi l)\{V_2(l) - V_2(0)\}^{1/2} \qquad (A4.19)$$

These techniques have been used to determine the parameters of a BM potential, an inverse power law, several screened Coulomb potentials and a Hartree potential derived from a self-consistent field calculation by Tucker *et al.* [7] and fit by a sum of three exponentials. Except in the case of a Firsov TF potential the screening radii obtained for (100) and (111) channels in Au, reproduced in Table A4.1 [2], are in good agreement with theoretical estimates and with other experimental results. The Firsov potential loses its validity rapidly as r increases (see Section 3.3) and will be excessively large for atoms traveling in the center of a channel: hence the very low experimental value for the screening radius.

The results of these channeling studies indicate that the technique is

TABLE A4.1[a]

DEFINITIONS AND ION-INDEPENDENT PARAMETERS OF SOME INTERATOMIC
POTENTIAL FUNCTIONS BASED ON EXPERIMENTAL CURVATURE
PARAMETERS [5] FOR SEVERAL IONS IN Au

Potential (abbreviation)	Definition of $V(r)$	Ion-independent parameter[b]
Power	a_k/r^k	$k = 3.50$
Born–Mayer (BM)	$C_{BM} \exp(-r/a_{BM})$	$a_{BM} = 0.230$ Å
Screened Coulomb	$(Z_1 Z_2 e^2/r)\phi(r)$	
Power (PS)	$\phi(r) = 1 - r[r^m + c^m]^{-1/m}$	$m > 2.50$
Thomas–Fermi (TF)	$\phi(r) = \phi_{TF}(r/a_{TF})$	$a_{TF} = 0.0172$ Å
Bohr (B)	$\phi(r) = C_B \exp(-br)$	
Exponential sum	$\phi(r) = \sum_{i=1}^{3} \alpha_i \exp(-\beta_i r)$	
Molière (M)	$\begin{cases} [\alpha] = [0.35, 0.55, 0.10] \\ [\beta] = [b, 4b, 20b] \end{cases}$	$b = 3.22$ Å^{-1} $1/b = 0.310$ Å
Hartree (H) (Au only)	$\begin{cases} [\alpha] = [0.25, 0.30, 0.25] \\ [\beta] = [b, 2.43b, 8.77b] \end{cases}$	

[a] After Robinson [2].
[b] Based on the experimental value $y_{111}/y_{100} = 0.604$; $l_{111} = 1.1774$ Å; $l_{100} = 1.0197$ Å.

excellent for obtaining the interaction potential provided that the channeled
ion consists of a weakly screened nucleus. Good agreement was obtained for
alpha-particles but much poorer for iodine ions. This is to be expected since
the screening of an iodine ion will vary with its position relative to the walls
of the channel. Nevertheless, there is good evidence that the method may be
used with some confidence to probe the one-atom potential of a variety of
crystalline materials.

REFERENCES

1. M. T. Robinson, *Phys. Rev.* **179**, 327 (1969).
2. M. T. Robinson, *Phys. Rev.* **B4**, 1461 (1971).
3. S. Datz, C. D. Moak, T. S. Noggle, B. R. Appleton, and H. O. Lutz, *Phys. Rev.* **179**, 315 (1969).
4. S. Datz, C. D. Moak, B. R. Appleton, M. T. Robinson, and O. S. Oen, "Atomic Collisions in Solids," p. 374. North-Holland Publ., Amsterdam, 1970.
5. B. R. Appleton, S. Datz, C. D. Moak, and M. T. Robinson, *Phys. Rev.* **B4**, 1452 (1971).
6. F. H. Eisen and M. T. Robinson, *Phys. Rev.* **B4**, 1457 (1971).
7. T. C. Tucker, C. D. Roberts, C. W. Nestor, Jr., and T. A. Carlson, *Phys. Rev.* **178**, 998 (1969).

AUTHOR INDEX

Numbers in parentheses are reference numbers and indicate that an author's work is referred to, although his name is not cited in the text. Numbers in italics show the page on which the complete reference is listed.

A

Abarenkov, J. V., 93, 101, *110*, 115, *144*
Abrahamson, A. A., 40 (4, 5), 41 (6, 7), 42, 46, 47, *48*
Afrosimov, V. V., 181 (22, 23), 185, 186 (23), *189*
Amdur, I., 176 (1), 177 (4, 5, 6, 7, 8), 180, *189*
Anderson, N., 31, 32 (32), *34*
Animalu, A. O. E., 95, *110*, 116, 117 (6, 8), 120, *144*
Appleton, B. R., 232 (3, 4, 5), *236*
Arthurs, A. M., 31, 32 (32), *34*
Ascarelli, P., 196, 197, *207*

Ashcroft, N. W., 96, *111*, 121, 122, 123, *145*, 198, 199, *207*

B

Baker, E. B., 26 (9), *34*
Bardeen, J., 98, *111*
Baroody, E. M., 161 (7), *170*
Barr, L. W., 230 (7), *231*
Bauer, W., 80 (36), 81 (36), *87*
Beeler, J. R., Jr., 211 (8), 214 (17), 219 (26, 27), *222*
Berry, H. W., 178 (9, 10), *189*
Besco, D. G., 219 (26), *222*

237

SUBJECT INDEX

A

Abrahamson potentials, 40–47
Alkali metals, 106, 117, 120, 121, 124–129, 196
Alloy potentials, 133–137
Angular momentum, 150, *see also* Quantum numbers
Annealing, 218
Argon potentials, 39
  Abrahamson, 41, 47
  Bohr, 165
  experimental, 165, 177, 183
  Firsov, 53, 165
  Lennard-Jones, 53–54
Asymptotic potential, 105, 131, 195
Atom–atom scattering experiment, 177, 188
Atomic beams, 147, 172–176, 188

Atomic collisions, 13, 14
  analytical potentials for, 57–19
  theory of elastic, 146–170
  see also Atom–atom scattering experiments, Ion–atom scattering experiments, Radiation damage
Atomic displacement cascade, *see* Displacement cascade

B

Band structure energy, 103, 110, 137, 143
Bare ion potential, 94–98, 107
  Ashcroft, 96, 97, 121–126
  Heine–Abarenkov–Animalu (HAA), 94, 95, 115–121, 125
  Ho, 97, 126–129
  point ion, 98, 135
  Shaw, 95, 96, 119